Phytochemistry and Angiosperm Phylogeny

Edited by
David A. Young,
David S. Seigler

PRAEGER

PRAEGER SPECIAL STUDIES • PRAEGER SCIENTIFIC

Library of Congress Cataloging in Publication Data
Main entry under title:

Phytochemistry and angiosperm phylogeny.

"A compilation of the papers presented at the symposium 'Phytochemistry and angiosperm phylogeny' held on August 15, 1979, at Oklahoma State University, Stillwater, under the primary sponsorship of the Phytochemical Section of the Botanical Society of America as part of the 30th Annual Meeting of the American Institute of Biological Sciences"--Introd.
   Contents: DNA hybridization techniques for the study of plant evolution / H.S. Belford, W.F. Thompson, and D.B. Stein -- Amino acid sequence studies and plant phylogeny / Ron Scogin -- Divergence, convergence, and parallelism in phytochemical characters / James Eric Rodman -- [etc.]
   1. Angiosperms--Congresses. 2. Phylogeny (Botany)--Congresses. 3. Plant chemotaxonomy--Congresses. 4. Botanical chemistry--Congresses. I. Young, David A. II. Seigler, David S. [DNLM: 1. Biochemistry--Congresses. 2. Phylogeny--Congresses. 3. Plants--Congresses. 4. Plants--Classification--Congresses. QK 495 P578 1979]
QK495.A1P48   582.13'0438   81-8603
 ISBN 0-03-056079-9   AACR2

Published in 1981 by Praeger Publishers
CBS Educational and Professional Publishing
A Division of CBS, Inc.
521 Fifth Avenue, New York, New York 10175 U.S.A.

© 1981 by Praeger Publishers

All rights reserved

123456789 145 987654321

Printed in the United States of America

LIST OF CONTRIBUTORS

H. S. Belford, Department of Plant Biology, Carnegie Institution of Washington, Stanford, California 94305

Rolf M. T. Dahlgren, Botanical Museum of the University of Copenhagen, Gothersgade 130, DK 1123 Copenhagen, Denmark

Bent Juhl Nielsen, Institute of Organic Chemistry, The Technical University, DK 2800 Lyngby, Denmark

James Eric Rodman, Department of Biology, Yale University, New Haven, Connecticut 06520

Soren Rosenthal-Jensen, Institute of Organic Chemistry, The Technical University, DK 2800 Lyngby, Denmark

Stephen G. Saupe, Department of Botany, University of Illinois, Urbana, Illinois 61801

Ron Scogin, Rancho Santa Ana Botanic Garden, Claremont, California 91711

David S. Seigler, Department of Botany, University of Illinois, Urbana, Illinois 61801

Dale M. Smith, Department of Biological Sciences, University of California, Santa Barbara, California 93106

D. B. Stein, Department of Plant Biology, Carnegie Institution of Washington, Stanford, California 94305

W. F. Thompson, Department of Plant Biology, Carnegie Institution of Washington, Stanford, California 94305

Robert F. Thorne, Rancho Santa Ana Botanic Gardens, Claremont, California 91711

David A. Young, Department of Botany, University of Illinois, Urbana, Illinois 61801

CONTENTS

List of Contributors   V
Introduction   VIII

DNA Hybridization Techniques for the Study of
Plant Evolution
    H. S. Belford, W. F. Thompson and D. B. Stein   1

Amino Acid Sequence Studies and Plant Phylogeny
    Ron Scogin   19

Divergence, Convergence, and Parallelism in Phytochemical Characters: The Glucosinolate-Myrosinase System
    James Eric Rodman   43

Cyanogenic Compounds and Angiosperm Phylogeny
    Stephen G. Saupe   80

Terpenes and Plant Phylogeny
    David S. Seigler   117

A Revised Classification of the Angiosperms with Comments on Correlation Between Chemical and Other Characters
    Rolf M. T. Dahlgren, Søren Rosendal-Jensen and Bent Juhl Nielsen   149

The Usefulness of Flavonoids in Angiosperm Phylogeny: Some Selected Examples
    David A. Young   205

Phytochemistry and Angiosperm Phylogeny: A Summary Statement
    Robert F. Thorne   233

INTRODUCTION

Dale M. Smith

Department of Biological Sciences
University of California
Santa Barbara, California 93106

This volume is a compilation of the papers presented at the symposium "Phytochemistry and Angiosperm Phylogeny" held on August 15, 1979, at Oklahoma State University, Stillwater, under the primary sponsorship of the Phytochemical section of the Botanical Society of America as part of the 30th Annual Meeting of the American Institute of Biological Sciences. Co-sponsorship came from the American Society of Plant Taxonomists and the Physiological section of the Botanical Society of America. Dr. David A. Young and Dr. David S. Seigler were primarily responsible for the Symposium organization and presentation. Support of the Symposium by the National Science Foundation is gratefully acknowledged.

Taxonomic botany in particular, having been born of the shaman's art and the herbalist's practice of medicine, should be most benefited by a renewed occupation with the chemical properties of plants. As if it were necessary, the truths of molecular genetics validate early reliance upon chemical data in plant classification and reconfirm our present desire to see a perfect consummation of the relationship of botany and chemistry. Thus, it is perfectly appropriate that phytochemists and systematists should pool their insights to discuss the role of chemistry in the study of phylogeny.

Rapid advances have been made in the methodology of plant chemical analysis. Flavonoids continue to be the chemical substances that are at the core of most comparative phytochemical studies, and their identification has been almost routine in many systematics laboratories. Rapid chromatographic and spectroscopic methods suggest the possibilty of automation of these procedures in the near future. Ultimately, lists of flavonoids and many other constituents may be as common as tables of leaf shapes and trichome types. In view of the likelihood of this information becoming easily available, it is imperative that rationale and procedures be developed to incorporate it with greatest facility into our arsenal of weapons for attacking the question of angiosperm phylogeny. Symposia such as these are exceedingly important in aiding this development.

Yet, in spite of the evidence that changes in chemical organization are at the root of evolutionary change, acceptance of chemical characters as evidence to be used in phylogenetic studies has not been universal. Among those who have accepted chemical characters as valid in phylogenetic studies, the level of acceptance has varied. Some reluctantly include chemical characters when they confirm conclusions reached from other directions, while at the opposite extreme it is contended that the ultimate success in such an endeavor merely awaits the arrival of the proper level of chemical sophistication. Thus, a real problem exists for even the staunchest supporter of the chemical approach in deciding which chemical data one should use, how they should be ranked with other biological facts, and in what context these data should be employed.

Even though the dogma of molecular biology holds that chemical characters have a genetic basis, and thus an antecedent-descendent relationship, it must be borne in mind that the _exact_ mode of inheritance and the _exact_ biochemical pathways have been shown for relatively few _natural_ substances in only a small sampling of species (although many families have been sampled to a limited extent). In many plant groups, chemical knowledge is only in the folklore stage. Major problems in interpretation and utilization of chemical data will exist and many controversies will arise until the more important knowledge gaps are bridged.

This symposium was organized to bring together phytochemists and plant systematists to pool their knowledge in the area of angiosperm phylogeny. Participants were chosen to present assessments of the current status and contributions of the newest approaches as well as summaries of more well-established methodologies. It was also deemed advisable to show how group-specific chemicals have systematic value in a narrow context and also how generally-distributed substances have systematic value of a different sort; thus, both specialists and generalists participated. Contributions were specifically solicited from young scientists whose viewpoints are expressed alongside those of established scientists whose viewpoints are more widely known. Divergent opinions are also expressed by the contributors and this was the goal of the symposium.

DNA HYBRIDIZATION TECHNIQUES FOR THE STUDY OF PLANT EVOLUTION[1]

H. S. Belford[2], W. F. Thompson[3],
and D. B. Stein[4]

Department of Plant Biology
Carnegie Institution of Washington
Stanford, California 94305

Both qualitative and quantitative questions may be asked by systematists. For example, "Are peas more closely related to beans, to beets or to snapdragons?" Or "How closely, or to what extent, are peas related to beans?" Ideally, we would like to turn to fossil evidence to answer these questions. However, at present the fossil record for angiosperms is somewhat limited. Therefore, the classical approach to these systematic questions has been to catalog physical attributes, assign subjective character weights and tally up scores. While these methods are certainly useful, they do suffer from definite limitations such as the necessity of character weighting or interpretation of primitive versus derived character status. Meanwhile, hidden in every organism there is an evolutionary record for the taking-- the genomic DNA sequences passed from generation to generation. It is this genetic record which systematists have begun to exploit in analyzing the results of sexual hybridizations and in looking for similarities among species in chromosome number, size and banding patterns. It is here that molecular biology intersects with the study of systematics.

---

[1] Publication No. 692 from the Department of Plant Biology, Carnegie Institution of Washington. Supported in part by NSF Grant DEB 76-83405 (to W.F.T.).

[2] Present address: Department of Biochemistry, School of Medicine, Boston University, Boston, Massachusetts 02118

[3] Also, Department of Biology, Stanford University.

[4] Present address: Department of Zoology, University of Massachusetts, Amherst, Massachusetts 01002.

One of the most exciting developments in molecular biology over the last decade has been the concept of the molecular clock (for reviews see Wilson, et. al., 1977; Fitch 1976; and Kohne 1970). A combination of data from the fossil record and molecular studies has been used to show that single nucleotide substitutions tend to accumulate at uniform average rates during evolution. That is, the rate of change in the base sequences, or primary structure, of DNA exhibits clock-like behavior, steadily incorporating point mutations quite independently of bursts and lulls in phenotypic and chromosomal evolution. There is also evidence for protein clocks (see Scogin, this volume). However, it is our belief that the DNA clock is more easily exploited for the study of systematics than protein clocks because the DNA complement contains not only sequences specifying proteins but also a large array of sequences with no known function. This "secondary" DNA may include more than 90% of the total DNA complement (even in the "single copy" fraction). It appears to be under minimal selective constraint regarding its base sequence and thus to evolve at a uniform and perhaps more rapid average rate than most proteins (reviewed by Britten and Davidson 1976; Hinegardner 1976).

Thus, by measuring differences in the primary structure of the DNA in modern species, it has become possible to estimate the relative amount of time elapsed since their divergence from ancestral lines. Where fossil histories are available for any one pair of species in a phylogenetic study, relative divergence time estimates may be converted to geologic time units for all species pairs. This has already been done for some species of sea urchins (Angerer, et al. 1976; Harpold and Craig 1978), amphibians (Galau et al. 1976), and mammals (Laird, et al. 1969; Kohne 1970; Fitch 1976). More recently, divergence rates have been estimated for three species in the fern genus *Osmunda* (Stein et al. 1979).

The advantages of DNA sequence divergence measurements over classical taxonomic approaches to phylogenetic problems are several. With hybridization the complete spectrum of DNA sequences present in a species is compared with that of other species, thus avoiding subjective decisions on which characters to use. Also, the hazard of basing a phylogeny on characters which might have been subject to great fluctuations in selective pressure (e.g., convergent evolution) is avoided. And finally, molecular approaches bypass the difficulty of prezygotic and postzygotic mating barriers.

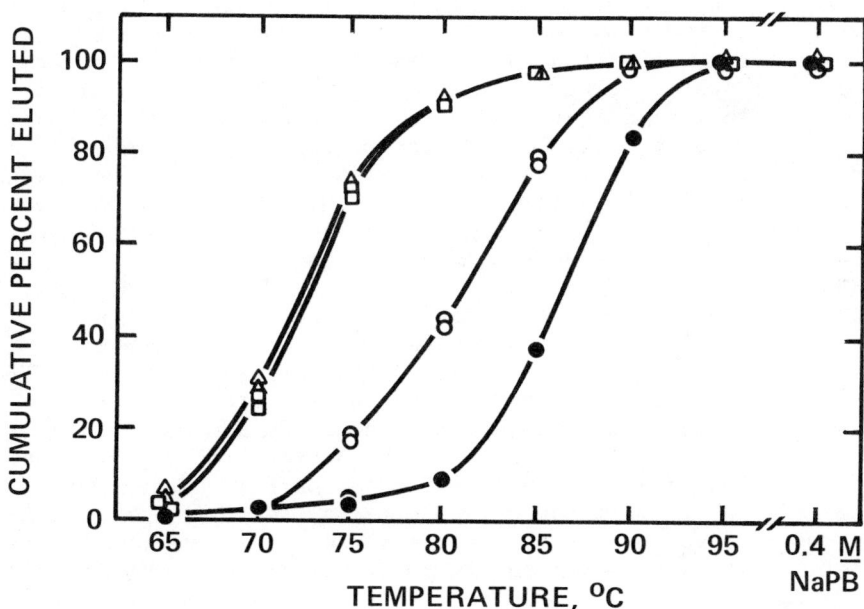

Figure 1. Thermal stability profiles of native and reassociated *Osmunda* species DNA. Native *O. claytoniana* DNA (●) ^{125}I-labelled *O. claytoniana* single copy DNA reassociated with unlabelled, total genome DNA of *O. claytonia* (o), of *O. cinnamomea* (□), and of *O. regalis* (▲). Native or reassociated DNA was bound to a column of hydroxyapatite (HAP) at 60°C in 0.12M sodium phosphate (pH 7.0) and subsequently eluted in similar buffer by raising the column temperature in increments of 5°C. See text for further explanation. (From Stein, et al. 1979).

MEASURING NUCLEOTIDE BASE SEQUENCE DIVERGENCE

Sheared fragments of DNA can be denatured into single strands by heat or alkali treatment. Subsequently, complementary single strands can reform duplex structures when inclubated under carefully controlled conditions. The rate and precision of reassociation is dependent upon temperature, salt concentration, and the degree of strand complementarity. DNA fragments purified from two species can be mixed, denatured and then permitted to renature (or hybridize) under conditions dictating that only sequences with a high degree of base sequence homology (at least 75-80%) can pair. Following DNA hybridization, the reaction mixture is assayed for two characteristics: (1) the percentage of each DNA which is able to hybridize with sequences of the other species, and (2) the degree of base sequence divergence among those sequences able to form interspecific hybrids.

Details of the techniques involved in these measurements have been thoroughly reviewed elsewhere (e.g., Britten et al. 1974; Stein et al. 1979). Briefly, the extent of interspecific hybridization is assayed by passing the reaction mixture over a column of hydroxyapatite at $60°C$ in 0.12 M sodium phosphate buffer (pH 7.0), conditions in which double-stranded, reassociated DNA is bound while single-stranded, unreassociated DNA is not. The results are commonly plotted as the fraction of DNA reassociated versus C_0t (DNA concentration in moles nucleotides/liter x time in seconds; see Fig. 3). The extent of base pair mismatch (sequence divergence) within hybrid duplexes can then be determined by thermal denaturation. Hybridized sequences are again bound to hydroxyapatite. The temperature of the hydroxyapatite column is subsequently raised in steps (e.g., from $60°C$ to $100°C$ in $5°C$ increments). DNA which has been thermally denatured into single-strands is eluted from the column at each step. Figure 1 illustrates such thermal denaturation profiles for hybrid duplexes among three species of fern genus *Osmunda* (Stein et al. 1979). The temperature at which 50% of the originally bound DNA has eluted is called the T_E or average thermal elution temperature. The difference between the T_E values for interspecific (heterologous) and intraspecific (homologous) duplexes is directly proportional to the base sequence divergence between species. In the case of *Osmunda* species, both interspecific hybrids (*O. regalis* L. x *O. claytoniana* L. and *O. cinnamomea* L. x *O. claytoniana*) have ΔT_E values about $8.5°C$ below that of the homologous *O. claytoniana* duplexes. Thus, the base sequences of *O. regalis* and *O. cinnamomea* are equally diverged (here ca. 8.5%) from the sequences of *O. claytoniana*. From these data it may be inferred that *O. regalis* and *O. cinnamomea* became generally isolated from *O. claytoniana* at about the same time.

Angiosperms, like all eukaryotes thus far examined, contain two types of DNA sequences: single copy sequences, present only once per haploid genome; and repeated sequences, which may be represented in as many as a million copies. The repeated sequences arise in unpredictable amplification events and complicate the measurement of species divergence. Point mutations can occur in all DNA sequences, including those belonging to families of repeated DNA. Consequently, if the hybridization products of repeated sequences are compared between species, any changes in sequence which have occurred since the divergence of the species may be obscured by the longer history of change within the repeated families. Therefore, interspecific divergence time estimates are best conducted with single copy sequences.

The general procedure for the isolation of single copy sequences is as follows: Total genome DNA is denatured and then reacted to a point at which all repeated sequences are reassociated, but at which time single copy sequences, present in relatively lesser concentration, are substantially unreassociated and therefore separable from repeated sequences by hydroxyapatite fractionation. When repeated sequences comprise a relatively small fraction of the total genome (as they do in most animals), the single copy isolation procedure is straightforward. However, in many plants repeated sequences make up 60-90% of the genome, and single copy isolation can be difficult. Frequently, several cycles of reassociation and fractionation are required. For some plants, such as peas, the abundance of highly divergent repetitive sequences may make it essentially impossible to isolate very much pure single copy DNA (Murray, et al. 1978; Thompson and Murray 1979). Because of the difficulties in isolating plant single copy sequences and in the quantitative interpretation of measurements based on repeated sequences, considerable care must be taken to thoroughly characterize the DNA used in plant divergence studies.

SEQUENCE DIVERGENCE MEASUREMENTS IN ANGIOSPERMS

For specific illustrations of base sequence divergence measurements in angiosperms, we draw upon our work exploring the phylogenetic relationships among eight species of the genus *Atriplex* (Chenopodiaceae) using single copy sequence hybridization (Belford and Thompson 1979 and in preparation).

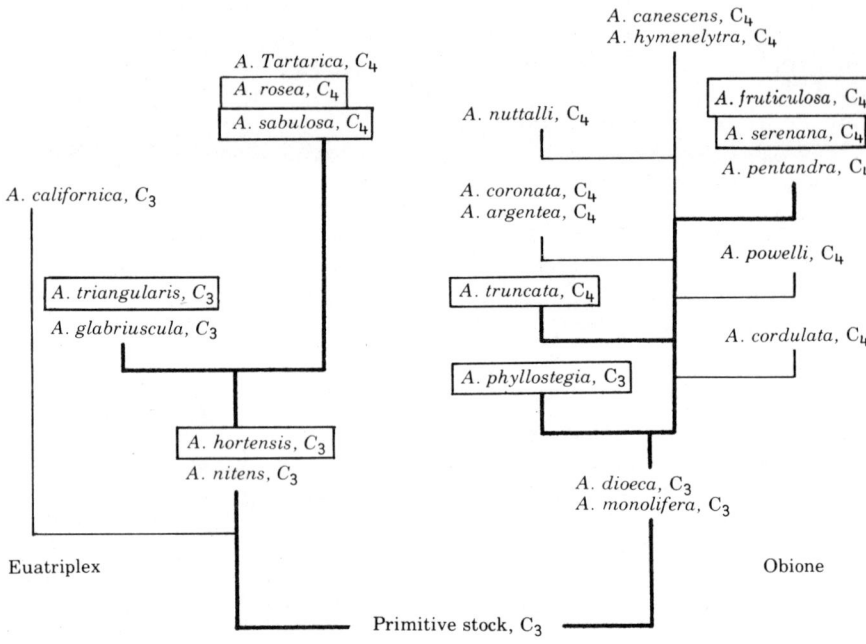

Figure 2. Phylogenetic scheme for *Atriplex* based on morphological criteria. This figure is adapted from Figure 29 of Hall and Clements (1923) monograph of *Atriplex*. C_3 and C_4 designations indicate the photosynthetic carbon fixation pathway used by each species as determined by Björkman, et al. (1973). The species in boxes are those between which single copy DNA sequences comparisons were made. The greater the distance of a species from the "primitive stock", the fewer are the "primitive" traits displayed by the species.

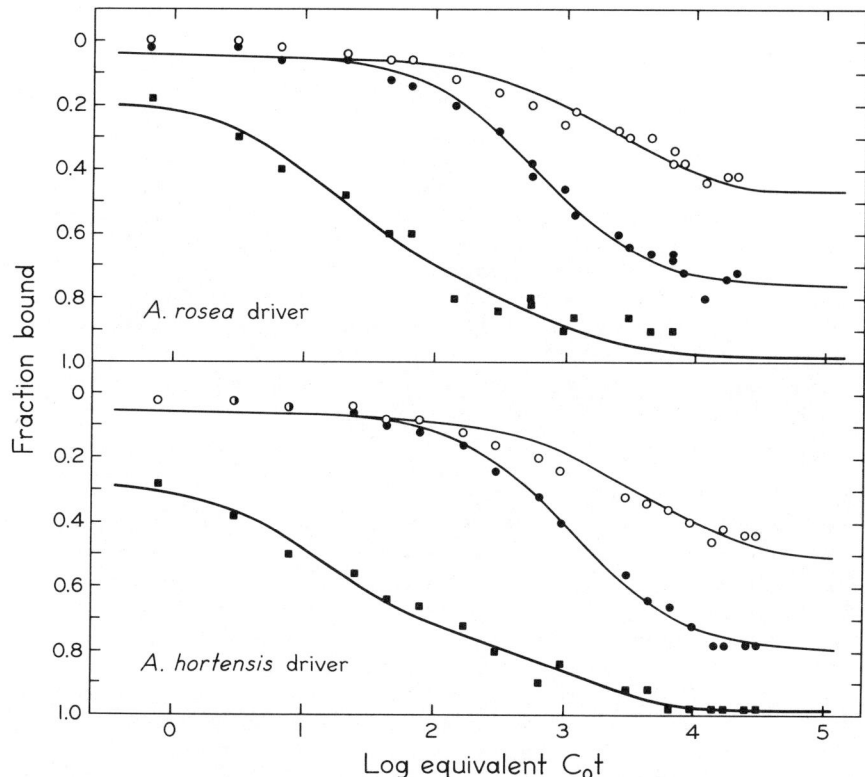

Figure 3. Reassociation kinetics for *A. rosea* and *A. hortensis* single copy and total genome DNA sequences. All DNAs were reassociated in 0.6M sodium phsophate (pH 7.0), 0.1 mM Na$_2$ EDTA at 66°C, then assayed by HAP fractionation at 60°C. C_ot values have been corrected to standard conditions. The reassociation data of total genome DNA (■) and homologous single copy sequences with total DNA (●) were fitted by a computerized least squares method using one or two ideal second order components. Reassociation data for heterlogous single copy sequences reacted with total genome DNA (●) were fitted by an equation modified to take into account the retarded rate at which heteroduplexes form as a result of base sequence mismatch (Galau, et al., 1976). Top panel: reassociation of *A. rosea* total genome DNA with itself (■), with *A. rosea* single copy sequences (●), and with *A. hortensis* single copy sequences (○). Bottom panel: reassociation of *A. hortensis* total genome DNA with itself (■), with *A. hortensis* single copy sequences (●), and with *A. rosea* single copy sequences (○). Note that the data have not been normalized to remove portions of the DNA unable to reassociate under any circumstances (1-5% of total genome DNA; ca. 20% of *in vitro* labeled single copy tracer preparations).

Atriplex is an interesting genus in that morphological and physiological data may be interpreted as suggesting that the C4 carbon fixation pathway had a polyphyletic origin within the genus. The phylogenetic hypothesis developed from the morphological studies of Hall and Clements (1923) is diagrammed in Figure 2, together with C_3 and C_4 designations from Björkman, et al. (1973). It would seem that C_4 species evolved from C_3 progenitors at least twice. Although such a polyphyletic origin is not impossible, it does seem somewhat improbable within a group of such closely related species. Therefore, a re-evaluation of phylogenetic relationships by means of DNA hybridization techniques was undertaken.

Preliminary experiments measured the extent of cross-reaction (the percentage of two genomes composed of similar sequences) with respect to two reference species, A. rosea L. and A. serenana A. Nels., chosen to represent the extremes of the morphologically-based phylogeny. Base sequence divergence was subsequently determined by thermal stability experiments. Cross-reactivity measurements reveal the fraction of the genome on which divergence time estimates are based, as well as providing an overall concept of how homologous two genomes are to each other. However, it must be recognized that cross-reactivity does not necessarily reflect divergence time because unpredictable delation events may reduce cross-reactivity in an unclocklike fashion. Figure 3 presents cross-reactivity results from one reciprocal pair of interspecific hybridizations. Purified and radioactively labeled single copy sequences from A. hortensis L. and A. rosea were hybridized with excess total DNA of A. rosea (top panel) or A. hortensis (bottom panel). Results from all Atriplex single copy cross-reactions are summarized in Table 1, together with values for spinach (Spinacia oleracea L.), also a member of the Chenopodiaceae). Reciprocal measurements have been made for most species combinations. The overall range of cross-reactivities is broad, rising from 10-15% in the case of the spinach-Atriplex combinations, to greater than 90% in the case of A. serenana-A. fruticulosa Jeps. A large number of cross-reactivities fall in the range of aout 50-60%. We interpret this reduced cross-reactivity to mean that extensive (and probably random) deletions have occurred during evolution of the various Atriplex genomes so that the modern species no longer share full cross-reactivity (i.e., many sequences lost in one species were not lost in a second species and vice versa). It is intriguing that as little as 30% of the single copy fraction (ca. 10% of the total genome) is required to specify characteristics common to all eight Atriplex species.

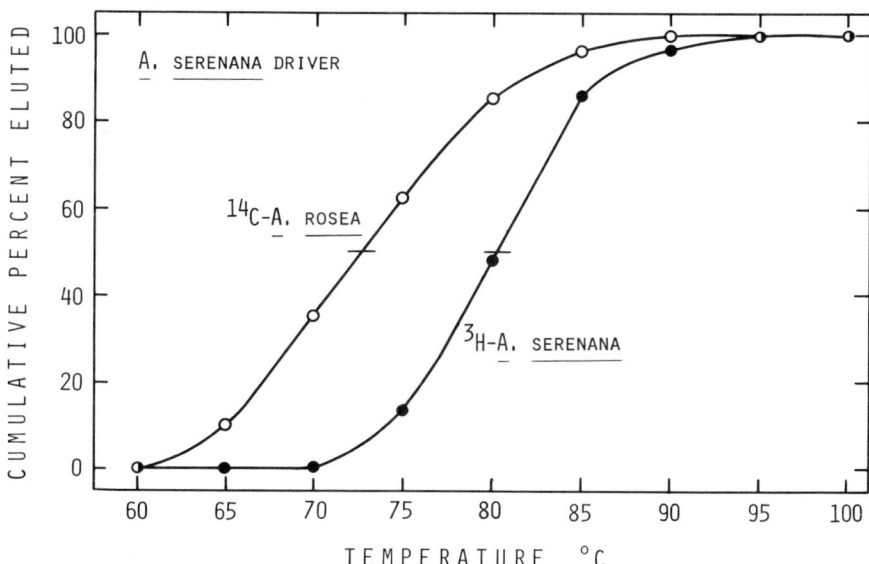

Figure 4. Thermal stability profiles of A. rosea and A. serenana single copy sequences hybridized with total DNA of A. serenan. ^{14}C-A. rosea and ^{3}H-A. serenana single copy sequences hybridized with unlabelled, total DNA of A. serenana, bound to HAP in 0.12 M sodium phosphate, 60 C, and thermally eluted in 5°C steps with 75 mM sodium phosphate buffer, under which conditions a 1°C ΔT_E is equivalent to approximately 2% base sequence divergence. Here, the T_E for the homologous ^{3}H-A. serenana hybrids is about 7.5°C above the heterologous ^{14}C-A. rosea-A. serenana hybrid T_E. Thus, A. serenana and A. rosea single copy sequences are approximately 15% diverged. The thermal stability profiles have been l corrected to remove any contribution from sequences other than single copy (Belford and Thompson, in preparation).

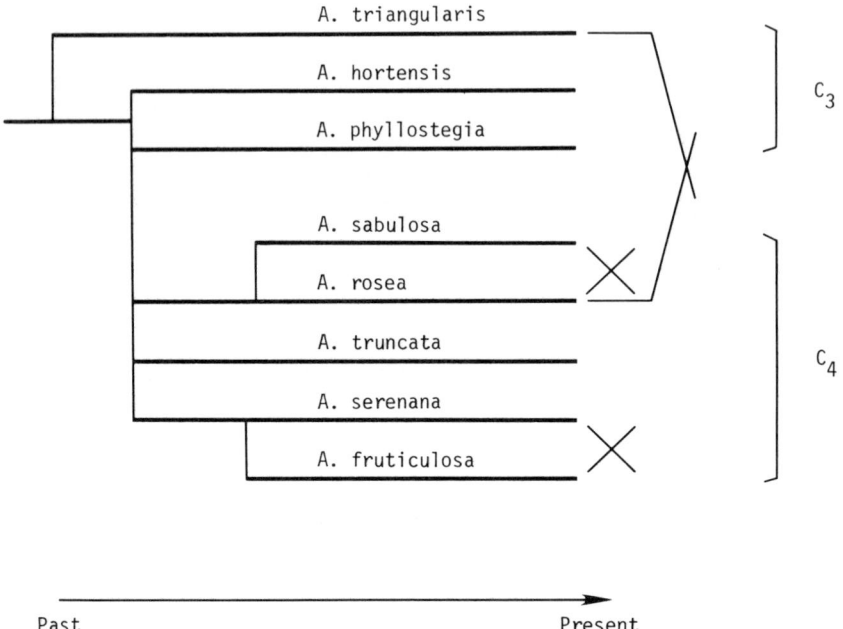

Figure 5. The phylogenetic scheme for *Atriplex* species based on interspecific nucleotide sequence divergence measurements. Horizontal distances from branch point to the present time are directly proportional to the time elapsed since the separation of species pairs within experimental error. The hypothesized secondary genetic exchange between *A. triangularis* and *A. rosea* (see text) is not shown, but would have occurred after the period of rapid speciation. Fossil evidence for the conversion of relative divergence times to geologic time is lacking for angiosperms. However, if an average divergence rate derived from animal sequence studies (Laird, et al. 1969; Fitch 1976; Angerer, et al. 1976; Galau, et al. 1976; Kohne 1970; Harpold and Craig 1978) is used, it is estimated that the period of rapid speciation occurred some 30-35 million years ago while the separation of the *A. rosea* and *A. serenana* lineages from *A. sabulosa* and *A. fruticulosa* lineages, respectively, occurred perhaps 20 million years ago. "Interfertility" indicates the successful sexual hybridizations (Nobs 1976). Photosynthetic pathways (C_3 or C_4) are indicated according to Björkman, et al. (1973).

Figure 4 shows the thermal stability profiles of duplexes formed between single copy sequences of *A. rosea* or *A. serenana* and total genome DNA of *A. serenana*. Interspecific hybrids show a T_E value approximately 7°C below the T_E of the homologous hybrids. Under our experimental conditions, a 7° ΔT_E indicates about 14% base sequence difference between the species (Belford and Thompson, in preparation). Table 2 summarizes thermal stability results for all species pairs, including reciprocal hybridizations. About one-third of the ΔT_E values in column 1 are the average of two separate experiments (experimental error was generally less than 1°C). Reciprocal measurements between *Atriplex* species were also equal within a 1-1.5°C error. However, the *Atriplex*-spinach ΔT_E results illustrate a problem that may limit the utility of the single copy sequence hybridization technique in plant systematic studies. In contrast to the situation for *Atriplex* species, these reciprocal ΔT_E values differ by as much as 4°C. This reciprocity failure probably stems from the difficulty of measuring accurate T_E values when cross-reactivity is very low. {Recall that *Atriplex*-spinach hybrids showed only 10-20% cross-reactivity (Table 1)}. If cross-reactivity above the genus level is frequently this low, quantitation of inter-generic comparisons will be quite difficult (see Summary and Perspectives).

It is interesting to note that the few *Atriplex*-spinach hybrids which do form are as stable as many of the *Atriplex*-*Atriplex* hybrids. This observation might be interpreted as suggesting that the *Atriplex*-spinach divergence event was as recent as some of the *Atriplex*-*Atriplex* divergence events. However, we feel it is much more likely that the reduced cross-reaction in these experiments involves a conserved fraction of sequences, evolving more slowly than the average total single copy sequence rate.

The lowest degree of base sequence divergence within *Atriplex* is seen in the *A. rosea* - *A. sabulosa* Jones and *A. serenana*-*A. fruticulosa* species pairs, which have ΔT_E values of about 4.0-4.5°C. Most of the other *Atriplex* pairs show slightly greater divergence, around 5.0-6.5°C. In comparable studies of animal genera, interspecific ΔT_E values are generally much less than observed in *Atriplex*. For example, no divergence is measureable among *Junco* species (Shields and Straus 1975), while the divergence among *Peromyscus* species (Rice 1972) is less than half as great as that among *Atriplex* species. From these data, limited as they are, one might infer

Table 1. Summary of cross-reactivity results. (Cross reactivity, or single copy homology between species, is listed both as the fraction and as the amount (nucleotide pairs or NTP) of single copy DNA which is able to hybridize between species. In the first two columns of data, single copy sequences from species A and B have been reacted with total DNA of species A. The two columns on the right summarize cross-reactivity in the reciprocal reaction, in which the two single copy fractions are reacted with total DNA from species B. The amount of DNA is based on experimentally determined genome sizes and single copy complexities (Belford and Thompson, in preparation). Genome sizes varied across a two-fold range, from 3.5×10^8 - 7.3×10^8 NTP, for *Atriplex* species. The spinach genome was slightly larger at 7.9×10^8 NTP. Single copy sequences made up to 26% of the various genomes.)

| | | Cross Reactivity | | | |
| | | Driver A | | Driver B | |
Species A	Species B	Fraction	10^8 NTP	Fraction	10^8 NTP
A. rosea L.	*A. sabulosa* Jones	0.72	1.1	0.81	1.2
	A. triangularis Willd.	0.55 0.59	1.2	0.75	1.1
	A. hortensis L.	0.58	1.2	0.64	1.0
	A. fruticulosa Jeps.	0.55 0.56	0.8	0.60	0.9
	A. truncata (Torr.) Gray	0.53	1.1	0.62	0.9
	A. serenana A. Nels.	0.64 0.56 0.63	0.8	0.47 0.45	0.7
	A. phyllostegia (Torr.) S. Wats.	0.26 0.33	0.6	0.26	0.4
	Spinacea oleracea L. (spinach)	0.12	0.2	0.21	0.3
A. serenana A. Nels.	*A. fruticulosa* Jeps.	0.96 0.89	1.3	----	----
	A. sabulosa Jones	0.81	1.3	----	----
	A. truncata (Torr.) Gray	0.72	1.5	0.93	1.2
	A. triangularis Willd.	0.56	1.4	----	----
	A. hortensis L.	0.46	0.9	----	----
	A. rosea L.	0.47 0.45	0.7	0.64 0.56 0.63	0.8
	A. phyllostegia (Torr.) Watts	0.28	0.5	0.41	0.5
	Spinacea oleracea L. (spinach)	0.18	0.4	0.22	0.3

that plant genera such as *Atriplex* may include species with longer evolutionary histories than do animal genera or, perhaps, that the molecular clock runs faster in plant genomes.

What sort of phylogenetic tree can we derive from these data? Figure 5 presents a scheme based on the relative divergence estimates in Table II. Horizontal distance is proportional to divergence. For example, the sum of the horizontal distances traceable between *A. fruticulosa* and *A. seranana* is proportional to the ΔT_E measured between the two species. Each branch point represents a divergence or speciation event. In contrast to the morphologically-based phylogenetic scheme of Fig. 2, speciation events in Figure 5 do not fall into a gradually reticulating pattern. Instead, assuming that base sequence divergence occurs at a relatively constant rate, it appears that five of the eight lineages studied originated in a single period of explosive speciation.

The *A. fruticulosa* and *A. serenana* lines diverged from each other at some time after this period, as did the *A. rosea* and *A. sabulosa* lines. The more recent nature of these events is supported by the genetic work of Malcolm Nobs, who showed that crosses between these species produced fertile F_1 progeny (Nobs 1976). The placement of *A. triangularis* Willd. with respect to the two reference species was not entirely clear. However, the large difference between *A. triangularis* and *A. serenana* suggests that the *A. triangularis* line may have diverged from the ancestral stock prior to the period of rapid speciation, while the low degree of base sequence divergence observed in experiments with *A. rosea* may be an indication that the *A. triangularis* line hybridized with the *A. rosea* line at some later time. Support for this secondary gene exchange hypothesis comes again from Nobs (1976), who was able to hybridize modern *A. triangularis* with *A. rosea* to produce a viable F_1 with at least partial fertility. Many other species combinations have been attempted with uniformly unsuccessful results.

The molecular phylogeny indicates to us that the subdivision of the *Atriplex* into two subgenera with primitive C3 and derived C4 species was probably artificial. Since all the C4 lineages we studied seem to have originated at about the same time (during the period of rapid speciation) it is no longer necessary to postulate more than a single origin for the C4 photosynthetic pathway within the genus.

Table 2. Summary of ΔT_E Measurements. In Column I, the ΔT_E values were obtained by comparing T_E values of single copy sequences of species A and B reassociated with total DNA of species A. In column II, the ΔT_E values were obtained from reciprocal reactions, that is, by comparing T_E values of species A and B single copy sequences reassociated with total DNA of species B.

| Species | | ΔT_E, °C | |
A	B	I	II
A. rosea	A. sabulosa	4.5	4.1
	A. triangularis	5.0	6.5
	A. truncata	5.2	6.5
	A. phyllostegia	5.5	4.3
	A. hortensis	6.2	5.4
	A. serenana	6.5	6.7
	A. fruticulosa	7.3	7.5
	Spinach	8.3	4.3
A. serenana	A. fruticulosa	4.4	---
	A. truncata	5.0	4.3
	A. sabulosa	5.0	---
	A. hortensis	5.3	---
	A. phyllostegia	5.8	4.8
	A. rosea	6.7	6.5
	A. triangularis	8.3	---
	Spinach	4.4	6.0

USE OF REPETITIVE DNA

One limitation inherent in the use of techniques such as those illustrated above is the degree of uncertainty introduced by experimental error in the ΔT_E measurements. Even a small uncertainty in ΔT_E might be equivalent to several million years of evolution. Thus, for example, we can say nothing concerning the actual order of divergence events for the five *Atriplex* lineages originating during the period of rapid speciation. These events may have been slightly separated in time, or they may have occurred virtually simultaneously. We believe that this type of uncertainty might be removed by means of careful comparisons of repeated sequence DNA families. As mentioned earlier, repeated sequences cannot be used to obtain quantitative estimates of divergence times, since they arise from unpredictable amplification and reamplification events which are not necessarily clocklike. However, amplification occurs frequently enough in evolution that the presence or absence of particular repeated sequence families in different genomes can often be used to distinguish different lineages.

This approach has been taken by Flavell and his colleagues (1977) in an impressive study of several cereal grains. The theory behind these experiments is straightforward. An amplification of some sequence or set of sequences in an ancestral line will be passed on to all daughter lines originating from the same ancestor. However, new amplifications occurring after separation of any two lineages will generally involve different sequences in each line, so that each new lineage has its own catalog of repeated sequences.

In the work by Flavell et al. (1977) the order of speciation events was determined for oats, barley, wheat and rye. Repeated sequences were hybridized between the four species to yield a matrix of repeated sequence cross-reactivities. Each of the four species genomes was shown to be composed of about 75% repeated sequences, while cross reactions ranged from 14 to 58%. Further experiments determined that once a repeated sequence group had arisen and been fixed into an ancestral line, it had remained without deletion.* By using

---

*Many deletion events would be required to eliminate all members of a typical repeated sequence family, so such families are expected to be much less sensitive to deletion than a single copy sequence would be.

the patterns of shared sequence groups, it was possible to construct a phylogeny specifying the order (although not the time) of divergence.

SUMMARY AND PERSPECTIVES

We have illustrated how existing techniques of DNA hybridization can be used to evaluate phylogenetic relationships in higher plants. Reactions involving only single copy sequences can be used for quantitative estimation of relative divergence times, while repeated sequence hybridization can sometimes provide additional qualitative information on the order of speciation events. However, there is still much to be done in order to develop these molecular techniques to their full potential. It is our belief that divergence time estimates based on single copy sequence hybridization will probably be especially helpful in sorting out plant relationships at or below the generic level. Above the generic level, cross-reactivities may be so reduced that divergence measurements might be difficult or inaccurate. This situation is in apparent contrast to that in most animals thus far examined. It may be that the evolutionary "turnover" rate of plant DNA (the balance between deletion and amplification events) is generally much faster than the "turnover" rate for animal DNA (Thompson 1978; Thompson and Murray 1979).

More distant plant relationships, at or above the genus level, may be most easily explored via repeat sequence comparisons because repeated sequences are less easily deleted from the genome (being present in greater number than single copy sequences). Along the same lines of reasoning, (i.e., using sequences less easily deleted) it may even be possible to derive divergence times among the more distantly related species by quantitating base sequence divergence in that small subset of single copy sequences acutally transcribed into mRNA. Such sequences would theoretically be more conserved and less easily deleted than non-transcribed sequences. This latter approach would somewhat resemble the phylogenetic technique based on protein amino acid sequences which is discussed elsewhere in this volume (R. Scogin), but would differ in that a broad spectrum of coding sequences would be compared simultaneously to obtain average divergence rates and distances. We conclude with the observation that it will indeed be interesting to learn to what extent, and with what result, plant phylogenetic relationships can be explored using these molecular techniques.

LITERATURE CITED

Angerer, R. C., E. H. Davidson and R. J. Britten. 1976. Single copy DNA and structural gene sequence relationships among four sea urchin species. Chromosoma 56:213-226.

Belford, H. S. and W. F. Thompson. 1979. Single copy DNA homologies and the phylogeny of *Atriplex*. Carnegie Inst. Wash. Year Book 78: in press.

Björkman, O., J. Troughton and M. Nobs. 1973. Photosynthesis in relation to leaf structure. Brookhaven Symp. Biol. 25: 206-226.

Britten, R. J. and E. H. Davidson. 1976. DNA sequence arrangement and preliminary evidence on its evolution. Fed. Proc. 35:2151-2157.

Britten, R. J., D. E. Graham and B. R. Neufield. 1974. Analysis of repeating DNA sequences by reassociation. Methods in Enzymology 29:363-418.

Fitch, W. M. 1976. "Molecular Evolutionary Clocks." In Molecular Evolution, edited by F. J. Ayala, pp. 160-178. Sunderland Massachusetts: Sinauer Associates.

Flavell, R. B., J. Rimpau and D. B. Smith. 1977. Repeated sequence DNA relationships in four cereal genomes. Chromosoma 63:205-222.

Galau, G. A., M. E. Chamberlin, B. R. Hough, R. J. Britten and E. F. Davidson. 1976. "Evolution of Repetitive and Nonrepetitive DNA." In Molecular Evolution, edited by F. J. Ayala, pp. 200-224. Sunderland Massachusetts: Sinauer Associates.

Hall, H. M. and F. E. Clements. 1923. The genus *Atriplex*. Carnegie Inst. Wash. Publ. 326:235-324.

Harpold, M. M. and S. P. Craig. 1978. The evolution of nonrepetitive DNA in sea urchins. Differentiation 10:7-11.

Hinegardner, R. 1976. "Evolution of Genome Size." In Molecular Evolution, edited by F. J. Ayala, pp. 179-199. Sunderland Massachusetts: Sinauer Associates.

Kohne, D. E. 1970. Evolution of higher organism DNA. Q. Rev. Biophys. 3:327-375.

Laird, C., B. L. McConaughy and B. J. McCarthy. 1969. Rate of fixation of nucleotide substitution in evolution. Nature 224:149-154.

Murray, M. G., R. E. Cuellar and W. F. Thompson. 1978. DNA sequence organization in the pea genome. Biochemistry 17: 5781-5790.

Nobs, M. A. 1976. Hybridization in *Atriplex*. Carnegie Inst. Wash. Year Book 75:421-423.

Rice, N. 1972. Changes in repeated DNA in evolution. Brookhaven Symp. Biol. 23:44-78.

Shields, G. F. and N. A. Straus. 1975. DNA-DNA hybridization studies of birds. Evolution 29:159-166.

Stein, D. B., W. F. Thompson and H. S. Belford. 1979. Studies on DNA sequences in the Osmundaceae. J. Mol. Evol.: in press.

Thompson, W. F. 1978. Perspectives on the evolution of plant DNA. Carnegie Inst. Wash. Year Book 77:310-316.

Thompson, W. F. and M. G. Murray. 1980. "Sequence Organization in Pea and Mung Bean DNA and a Model for Genome Evolution." In The Plant Genome, edited by D. R. Davies, in press. Norwich England: John Innes Institute.

Wilson, A. S., S. S. Carlson and T. J. White. 1977. Biochemical evolution. Ann. Rev. Biochem. 46:573-639.

AMINO ACID SEQUENCE STUDIES AND PLANT PHYLOGENY

Ron Scogin

Rancho Santa Ana
Botanic Garden

In historic perspective every approach to the solution of systematic questions seems to go through a characteristic progression of stages. Initially, there is euphoria among developers and adherents to a new and novel approach to the solution of taxonomic problems. They share the feeling that the touchstone for understanding systematic relationships has finally been found and that a panacea for taxonomic problems has been discovered. Subsequently, there follows heated discussion in various forums (usually generating more heat than light) as newfound proponents of the technique and its detractors both overstate their positions for emphasis. Then a period of reassessment sets in, during which the proper contribution of the new data source to systematics is critically analyzed and its value established. Finally, the new data is placed in proper perspective and incorporated into functioning systematic schemes. This progression of stages has characterized the introduction of new approaches to systematics during this century as illustrated first by cytology and cytogenetics, then biochemical systematics and, most recently, numerical approaches.

I wish to suggest at the outset that the use of amino acid sequence data in the study of plant phylogeny has currently progressed through euphoria and heated discussion and is now in a period of data reassessment and critical analysis, out of which will emerge an accurate judgement of the valid contribution of this data source.

The realization that genes and their primary products, amino acid sequences, contain historical information about origins of species emerged from the monumental discoveries of molecular biology during the 1950s. The realization grew that if one were seeking "living fossils", proteins were the place to look. Some proteins, like histones, have remained virtually unchanged since before the divergence of plants and animals. The first protein for which a sufficient number of sequences was compiled to conform homology among the sequences, and to warrant comparisons, was the mitochondrial respiratory protein, cytochrome c. Comparative studies of the early cytochrome sequence data culminated in the molecular, vertebrate phylogenetic tree presented by Fitch and Margoliash (1967). The striking agreement between the phylogenetic tree for vertebrates based on cytochrome c sequence comparisons and those based on the well-preserved and -documented vertebrate fossil record strongly reinforced the feeling that accurate phylogenetic trees could be constructed solely on the basis of amino acid sequence data from homologous proteins. These successes with vertebrates produced an optimistic hope that amino acid sequence data could be used to establish phylogenies in groups of organisms for which the fossil record is woefully inadequate, notably the angiosperms. In retrospect, it should be noted that the vertebrate phylogeny based on cytochrome c sequences is not without anomalies, specifically: (1) chickens appear more closely related to penguins than to ducks or pigeons; (2) man and monkey diverged from other mammals before the marsupial-placental (kangaroo-other mammals) divergence; and (3) turtles (reptile) appear more closely related to birds than to other reptiles (e.g., rattlesnake). However, at that early date, workers focused on the remarkable general agreement between the sequence data and the fossil record, rather than on the minor discrepancies.

THE USE OF AMINO ACID SEQUENCE DATA IN PLANT STUDIES

The first angiosperm cytochrome c sequence to be published was that of wheat (Stevens, et al. 1967), but it was not until the early 1970s that a sufficient number of plant cytochrome c sequences were determined in the laboratory of Professor Don Boulter at the University of Durham to warrant comparative studies. As plant amino acid sequence data accumulated through the efforts of Dr. Boulter and his collaborators, botanists, for the first time, had to consider a series of questions regarding the use of amino acid sequence data in systematics: (1) what is the attraction of comparative sequence data? (2) why should conclusions based on these data be considered valid? (3) what are the advantages and disadvantages of using amino acid sequence data over other approaches to systematic problems?

What is the Attraction of Amino Acid Sequence Data?

Amino acid sequence data is potentially attractive, especially to the plant systematist, for two reasons. Firstly, sequence data can provide a completely independent assessment of earlier phylogenetic schemes which are based almost totally on morphological features of extant taxa. Sequence data contain the potential to independently confirm existing schemes or allow resolution among mutually exclusive, conflicting schemes. Sequence data may also suggest previously unconsidered alternative relationships.

The second appeal of amino acid sequence data is that it may provide an objectively constructed phylogeny among taxa for which there is a poor fossil record (e.g., angiosperms).

Why Should the Conclusions Based on Sequence Data be Considered Valid?

The only reason for accepting the phylogenetic schemes based upon sequence data for plants is that it worked for vertebrates. As noted earlier, a phylogenetic tree based solely on cytochrome c sequences for vertebrates was in remarkable agreement with trees constructed from fossil evidence. There is no other "a priori" reason to expect it also to work for plants.

Confidence in the validity and accuracy of indicated relationships using only comparative sequence data increased (at least for lower taxonomic levels) with the publication of some early plant cytochrome c sequence data (Boulter, et al. 1972). Species which were closely associated (i.e., in the same family) on the basis of traditional morphological evidence exhibited very similar cytochrome c sequences. Therefore, amino acid sequence data satisfied the fundamental criterion for taxonomic utility; namely, correlation with other data sources (Cronquist 1976). The amino acid sequence data correlated with most of the morphological data upon which the close relationships of

certain taxa had been based. Examples of this correlation are
the two *Brassica* congeners (*B. napus* L. and *B. oleracea* L.)
which have identical cytochrome c sequences (Richardson, et al.
1971; Thompson, et al. 1971) and various genera occurring in
single, well-characterized families which cluster together in
sequence-based trees (e.g., *Helianthus* and *Guizotia* in the Asteraceae; *Gossypium* and *Abutilon* in the Malvaceae; and *Zea, Hordeum,* and *Triticum* in the Poaceae) (Boulter 1974). At least
for the lower taxonomic categories, sequence comparisons confirmed and reflected valid, universally accepted relationships.

What are the Advantages of Amino Acid Sequence Data Over Other Approaches?

There are several very attractive advantages to the use of
amino acid sequence data for phylogenetic studies. One was
noted above: namely, the possibility of construction of a
phylogenetic tree independent of traditional morphological data,
and thereby providing an independent test of those data.

A second advantage of sequence data lies in the possible
modes of analysis of these data. By virtue of their nature,
amino acid sequence data are highly amenable to sophisticated
quantitative analysis in a way that is not yet possible with
morphological characters. This analysis feature is derived
from our ever increasing fundamental understanding of the molecular nature of genetic information, the mechanism of mutation, and the fact that proteins are primary semantides or
direct readouts of segments of genetic information. Since an
amino acid sequence is a fixed readout of a nucleotide base
sequence, the number of sequence changing mutational events required among a collection of sequences can be quantified (given certain assumptions), and taxa containing those sequences
can be ranked in order of the numbers of mutations required
to generate one sequence from the other. Due to the complexity
of genetic control, epigenetic interactions, environmental influences, and probably other as yet undiscovered factors, this
type of analysis cannot be applied (at present) to morphological
data. The application of sophisticated quantitative analysis
to amino acid sequence data is, however, a double-edged sword.
Treatment of data in this manner is very complex and necessitates
whole new sets of assumptions based on mathematical tenets. The
very complexity of the mathematical data handling procedures
generates interpretive problems, which will be considered
later.

An additional advantage of sequence data is that they yield as a final product an objective, "non-intuitive" topology of relationships. This is in contrast to what is called by some as "eyeball taxonomy" or systematics by a bolt of lightning between the ears. Do not misunderstand my meaning: the human brain is a remarkable synthetic device for unconsciously correlating masses of data, and humans are strongly adapted to analyzing visual patterns. There is a definite place in systematics for intuitive insight based on experience. However, by contrast, sequence data yields an objective topology, constructed by fixed mathematical rules and procedures, and is based on clearly stated assumptions. Anyone following the same procedures should obtain the same result (it may be wrong, but it is reproducible!).

A final advantage to the use of sequence data lies in the possibility of the attachment of an absolute time scale to a phylogenetic tree. A time scale for trees based on sequence data is based upon the validity of current "molecular clock" models and must always ultimately be "calibrated" by reference to some fossil record, usually that for mammals. An early attempt to attach a time scale to angiosperm evolution based on cytochrome c sequence data was presented by Ramshaw, et al. (1972). This early effort was based upon the weak, but clearly and explicitly stated, assumption of a stochastic molecular clock in cytochrome c as reflected in a "unit evolutionary period" among mammals for clock calibration. The results of calculations based upon this assumption yielded a time of origin for angiosperms over 400 million years BP. That date of origin as several geological periods earlier than the oldest angiosperm fossils (from the early Cretaceous, ca. 125 million years BP) and the cytochrome-based time scale has not been accepted generally. Recent improvements in the model of the molecular clock (Fitch 1976) have demonstrated the weakness of the assumptions used by Ramshaw, et al. and now permit the calculation of a more generally acceptable time scale for angiosperm evolution based upon amino acid sequence data (Scogin, unpub. data). The attachment of an accurate time scale to molecular phylogenetic trees now appears to be a realistic goal of the near future, which is dependent only upon the collection of sufficient, appropriate sequence data.

What are the Disadvantages of Amino Acid Sequence Data?

There are several disadvantages to the use of amino acid sequence data in reconstructing phylogenetic histories. Some of these are purely practical or technical in nature, while others are conceptually fundamental.

A phylogenetic tree based upon a single protein is, in fact, an affinity tree of a gene, not a phylogenetic tree of biological species. This is derived from the fact that the tree was constructed from an analysis of the product of a single gene. Note is taken that in some cases several genes may be involved, as in the case of α- and β- chains of hemoglobin, but the "one gene-one polypeptide" hypothesis strictly limits the genome sampling of a tree based on sequences of a single protein. Therefore, compounding the technical and methodological difficulties of producing an accurate tree based on amino acid sequences is the fact that the tree produced is accurate for the sampled biological species only to the extent that the single gene studied is an adequate and accurate sampling of the total genome. The result is that even if a perfectly accurate gene affinity tree were constructed, it might not reflect the history of the total genome and therefore the biological species. We have no rigorous test at present of the degree to which a single gene is an adequate reflection of the complete genome. The "cutting edge" of natural selection acts upon organisms in populations and our understanding of the action of natural selection at the molecular level is far from perfect (Williams 1974). Thus, the selective history of the biological species could be very different from the selective history of a single gene.

The solution to this difficulty will require that sufficient sequences be known from two or more proteins in order to construct independent trees for each protein. The degree of congruence among these trees will be a measure of the degree to which the proteins reflect the total genomes, and have been subjected to equivalent selective pressures. A high degree of congruence among such trees will greatly increase the confidence level for sequence-based trees. A fact compounding the difficulty of overcoming this obstacle is that each protein has a fairly characteristic evolutionary rate and only proteins of approximately equal evolutionary rates can fruitfully be compared. To solve this problem, much time and effort will be required to accumulate sufficient numbers of sequences of different proteins with different evolutionary rates to permit adequate comparative testing.

A second disadvantage of amino acid sequence data can best be expressed as the lack of any fundamental advantage of these data over morphological data, with respect to detection and compensation for convergence. Convergence among morphological characters has long been the bane of traditional systematics, and it is no less a problem among amino acid sequences. The problem of convergence among amino acid sequences may be minimized by considering as many independent sequences as possible. Quantitative techniques are being developed to attempt to detect and correct for convergence at the data handling stage of sequence analysis. Peacock and Boulter (1975) have quantitatively examined, through computer simulation of evolutionary changes, the amount of expected error due to convergence under defined conditions, and have determined the properties of proteins and data handling techniques which minimize errors due to sequence convergence.

A third obstacle, at least temporarily, in using sequence data is the limited data set presently available. Many more sequences are known for animal proteins than for plant proteins. The number of known plant protein sequences is presented in Table 1. A further technical disadvantage of sequence data is the difficulty and expense of overcoming the obstacle of a limited data base. Obtaining amino acid sequence data is expensive in terms of hardware and time. Thus, new sequence data will continue to be generated mostly by established laboratories and, while technical advances may accelerate sequence data acquisition, the accumulation of useful numbers of sequences from new proteins will continue to be a slow process.

A final disadvantage to the use of amino acid sequences in systematics lies in the area of special problems associated with data handling. This will be dealt with at length in a later section.

RESULTS OF PLANT CYTOCHROME $\underline{C}$ STUDIES

Strategy

When the commitment of resources was made in the late 1960s to determine a sufficient number of plant cytochrome $\underline{c}$ sequences for comparative studies, certain technical constraints dictated a fundamental strategy. It was decided to study taxa across a wide taxonomic spectrum so that the results, when available,

Table 1. Amino Acid Sequence Data from Plants[1]

Protein	Taxonomic Group	Number of Sequences	
		Complete	Partial
Cytochrome c	Angiosperm	23	
	Gymnosperm	1	
Ferredoxin	Angiosperm	4	
	Fern Ally	1	1
Plastocyanin	Angiosperm	10	64
	Gymnosperm		3
	Ferns and allies		6
Papain	Angiosperm	1	
Bromelain	Angiosperm		1
Proteinase inhibitors	Angiosperm	8	5
Histone	Angiosperm	4	
Ribulose-1,5-diP Carboxylase	Angiosperm		6
Pollen Allergen	Angiosperm	1	
Concanavalin A	Angiosperm	1	
Monellin	Angiosperm	1	

[1]Data taken from Boulter (1978)

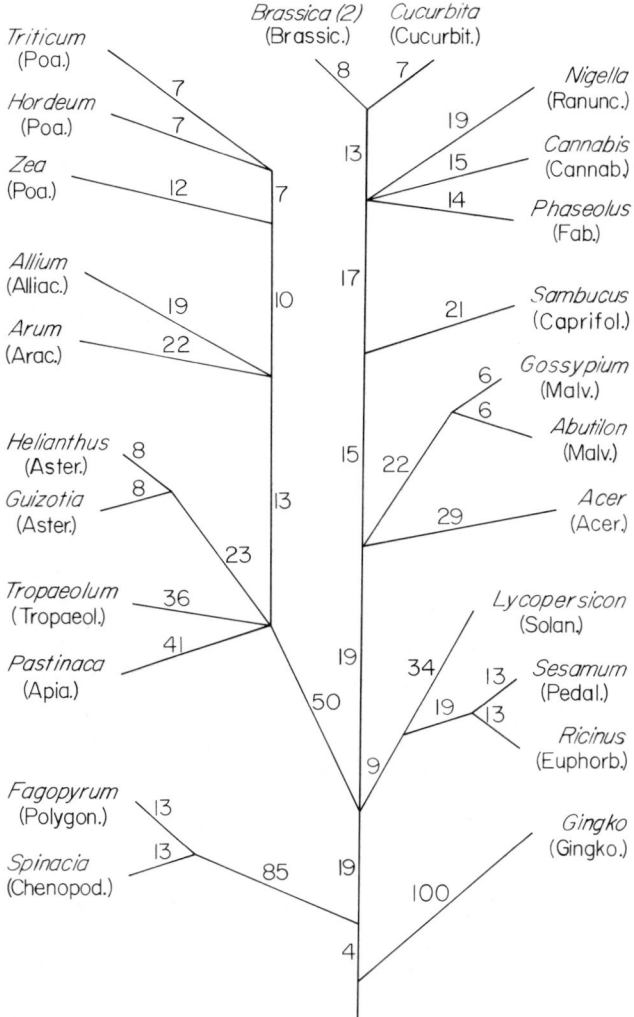

Figure 1. A phylogeny of higher plants based on cytochrome c sequences. Numerals refer to relative branch lengths based on a value of 100 for the distance of *Gingko* from the main evolutionary line axis. Redrawn from Boulter (1974).

would indicate general phylogenetic patterns among the angiosperms. Cytochrome c itself was chosen because of its ubiquitous occurrence among organisms and because of the large body of basic biochemical and sequence data already available for this protein from animal and fungal sources. The fact that the N-terminus of plant cytochromes c is acetylated precluded the use of automated sequencing techniques. About 15 mg of purified cytochrome c is required for sequence determination by manual Edman degradation with a redundant sequence determination for confirmation. In order to acquire that amount of protein, tissues with the highest cytochrome content (i.e., vigorously growing seedlings) had to be used. The hundreds of pounds of seeds necessary to provide sufficient cytochrome c dictated that, in general, horticultural and agricultural materials were examined.

The roadblock to rapidly accumulating sequences in these early days was not sequence determination per se, but rather purification of adequate amounts of sufficiently pure cytochrome c and the separation of complex mixtures of peptides resulting from enzymatic hydrolysis. By 1972 sufficient plant cytochrome c sequences had been determined and analyzed that Boulter, et al. (1972) presented a tentative phylogenetic tree for angiosperms based on amino acid sequences. That tree has been refined as additional sequences have become available (Boulter 1973, 1974). The most recently published cytochrome c phylogeny is shown in Figure 1.

The Cytochrome c Tree

The general pattern of angiosperm evolution which emerges from cytochrome c sequence studies is that of present day taxa (1) which represent the relictual ends of long and independent evolutionary histories and (2) the evolutionary common ancestors of which would not be referable to any contemporary taxonomic grouping (Boulter 1978). Alternatively phrased, sequence study results do not support schemes in which major contemporary plant groups are derived via currently recognized taxonomic antecedents in the sense of the Asteridae arising from the Magnoliidae via the Rosidae (Cronquist 1968). (For a dissenting view, see Cronquist 1976, p. 16).

Boulter, et al. (1972) have noted the prematurity of a detailed analysis of a cytochrome c phylogeny based upon such a very restricted data set, but the details of the cytochrome tree have been criticized by Cronquist (1976). While there is sufficient uncertainty regarding the details of phylogenetic placement in the cytochrome tree to make detailed consideration of individual families a less than productive activity, the general topology of the cytochrome tree appears to be a fairly accurate reflection of the protein's phylogeny (Peacock and Boulter 1975). There are, however, several, more general anomalies which are notable and warrant consideration. These are the age and affiliates of the Chenopodiiflorae (=Centrospermae) and the age and origins of the monocotyledons.

The evolutionary line leading to the Chenopodiiflorae (as represented among the sequence data by the Chenopodiaceae-*Spinacea*) appears to be the most ancient flowering plant yet examined to diverge from the general evolutionary line of the angiosperms. This divergence appears well before the divergence of the line leading to modern monocotyledons. Some support for the antiquity and isolation of the Centrospermaceous line is also found in the occurrence of unique betalain pigments among most members of this group (Mabry 1973) and protein inclusions in seive-tube plastids (Behnke and Turner 1971). In contrast, a large body of data concerning pollen morphology suggests that the Chenopodiiflorae are rather more recent in their origins and there is no fossil evidence for members of the group earlier than about 50 million years BP (Muller 1970).

The close affinity of the Polygonaceae with the Chenopodiiflorae has been traditionally accepted, but more recently questioned (Eckardt 1976). Cytochrome $\underline{c}$ data strongly support a close alliance between these two taxa.

Monocotyledons are generally accepted to have arisen very early in angiosperm evolution. This hypothesis is based largely on the occurrence of monosulcate pollen in the group (as well as several other morphological features). In sharp contrast, the results from cytochrome $\underline{c}$ sequence comparisons show the monocotyledons diverging almost contemporaneously with the Asteraceae. No rationalization of these two disparate models for the origin of the monocotyledons can be given at present. However, it should be noted that all monocotyledon sequences studied clustered together and were closely linked to both genera of the Asteraceae that were studied.

Sources of Error in the Cytochrome c Tree

If the cytochrome c sequence tree is flawed (and there is general agreement that it is, at least in its details), then why is it incorrect? A major source of possible error is that only 25 angiosperm sequences have been determined. This is certainly an inadequate sampling of the ca. 250,000 angiosperm species that are known (although only 17 animal cytochrome c sequences gave a stable and acceptable vertebrate tree). Further, these taxa are widely dispersed taxonomically, and as a result the fine resolution in tree topology is poor. As new sequences are added, some position assignments may shift slightly. A comparison of the cytochrome c trees of 1972 and 1974 reveal that the incorporation of 10 additional sequences yielded several significant changes of species position (Boulter, et al. 1972; Boulter 1974).

The major culprit in introducing distortions into the cytochrome tree remains undetected sequence convergences (i.e., similarities in sequence due not to descent from common ancestry, but rather to random chance, or an unidentified selective pressure resulting in the same residue in certain positions). Recent calculations have shown that approximately 33% of all amino acid substitutions within the cytochrome c data set are due to convergence (Peacock and Boulter 1975). Similar calculations for vertebrate cytochromes yield a somewhat lower value of 24% (D. Boulter, pers. com.). Data handling techniques which attempt to compensate for convergent subsitutions have been devised (Boulter, et al. 1979), but have not as yet been applied to the cytochrome c data set.

RESULTS OF PLASTOCYANIN STUDIES

Strategy

By 1974, the major thrust of sequence determination for systematic purposes in Professor Boulter's laboratory switched from cytochrome c to plastocyanin. Plastocyanin is a protein present in plant chloroplasts which participates in electron transport during photosynthesis. It is a copper-containing protein with a molecular weight of about 10,500 and, in higher plants, consists of 99 amino acid residues.

The initial strategy governing plastocyanin investigations was to determine numerous sequences across a wide taxonomic spectrum (similar to the cytochrome c strategy), and with these data construct a plastocyanin-based phylogenetic tree. The degree of congruence between the plastocyanin and cytochrome c trees would be a measure of the validity of using sequence data in phylogenetic studies, since if two independent proteins yielded similar results, confidence in the methods would be greatly enhanced.

The choice of plastocyanin was dictated by several considerations. The size of the molecule is appropriate for comparative studies and it is colored, greatly facilitating purification. More importantly, plastocyanin can be purified from leaf material in relatively high yields. It is present in leaves to the extent of about 6 mg of plastocyanin per kg of leaf material. Thus, sufficient quantities of plastocyanin for sequence determination could be prepared from virtually any systematicllly interesting plant for which several kg of leaf material were available. The N-terminal amino acid of plastocyanin is not blocked and thus automated sequencing techniques can be used.

As recently as 1975, Boulter and Ramshaw (1975) stated that the major constraints to progress in the use of amino acid sequence data in evolutionary studies lay in the purification of sufficient quantities of protein, and the subsequent separation of the complex mixtures of peptides resulting from the enzymatic hydrolysis necessary for manual sequence determination. These constraints were largely overcome by switching to a protein available in high yield from accessible materials and with the advent of methods for selectively coupling a particular peptide in a complex mixture to a solid support matrix (Laursen, et al. 1975). This support-bound peptide can be processed through a solid-phase, automated sequencing protocol which is capable of determining sequences at the rate of 20 residues per day.

The Durham research group has now determined 10 complete and 72 partial plastocyanin sequences from a wide variety of plant sources. However, difficulties have arisen when attempts were made to assemble these data into a general phylogenetic tree of angiosperms for comparison with the cytochrome c results. One problem is that plastocyanin evolves (i.e., fixes amino acid changing point mutations) at a rate 1 1/2 to 2 times as great as cytochrome c (Boulter, et al. 1977). Further, it is estimated that higher plant plastocyanin sequences may exhibit convergence in 40 to 50% of their residue positions (Peacock and Boulter 1975). These two factors (i.e., high rate of

Figure 2. An affinity tree among angiosperms based on plastocyanin sequence data. Members of single families are encircled. Numbers refer to species listed below. Redrawn from Boulter, et al. (1979).

1. *Ursinia anethoides*   Asteraceae
2. *Tussilago farfara*
3. *Senecio vulgaris*
4. *Bellis perennis*
5. *Hieracium* sp.
6. *Rudbeckia* sp.
7. *Cirsium vulgare*
8. *Phaseolus vulgaris*   Fabaceae
9. *Vigna radiata*
10. *Lupinus* sp.
11. *Vicia faba*
12. *Trifolium medium*
13. *Pisum sativum*
14. *Cytisus battendieri*
15. *Daviesia latifolia*
16. *Robinia pseudoacacia*
17. *Magnolia soulangeana*   Magnoliaceae
18. *Liriodendron tulipifera*
19. *Crataegus monogyna*   Rosaceae
20. *Prunus serrulata*
21. *Brassica oleracea* (cabbage) (a) Brassicaceae
21. *B. oleraceae* (cauliflower) (b)
22. *Capsella bursa-pastoris*
23. *Anthriscus sylvestris*   Apiaceae
24. *Heracleum sphondylium* (a)
24. *H. mantegazzianum* (b)
25. *Pastinaca sativa*
26. *Aegopodium podagraria*
27. *Solanum crispum*   Solanaceae
28. *Nicotiana tabacum*
29. *Capsicum frutescens*
30. *Solanum tuberosum*
31. *Lycopersicon esculentum*
32. *Lonicera periclymenum*   Caprifoliaceae
33. *Sambucus nigra*
34. *Viburnum tinus*
35. *Verbascum thapsus*
36. *Digitalis purpurea*
37. *Antirrhinum majus*
38. *Plantago major*   Plantaginaceae

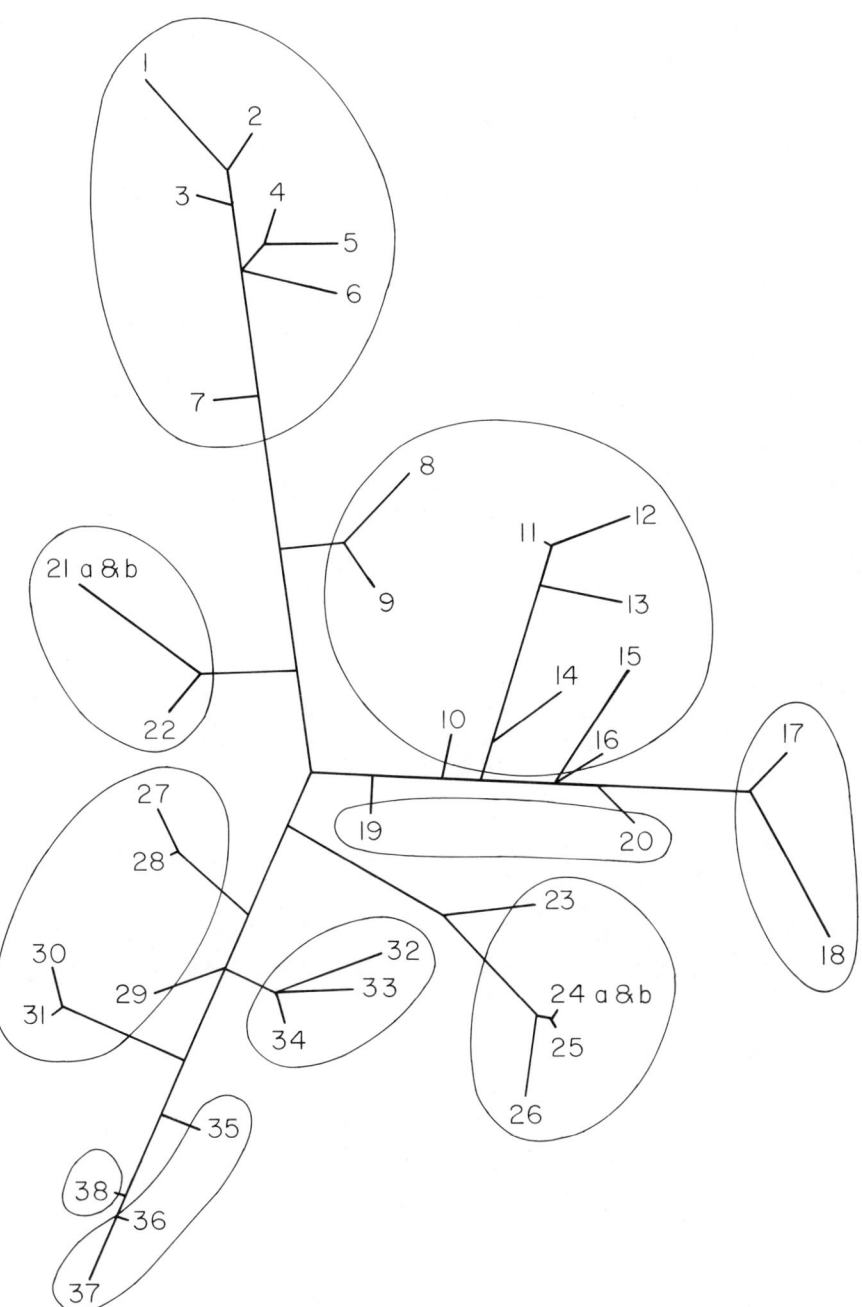

evolutionary change and substantial sequence convergence) combine to produce the situation that when attempts are made to generate trees over a wide taxonomic spectrum, many, equally likely, alternative trees are generated. Thus, plastocyanin sequences can, at present, be usefully employed only to determine relationships within a family or among closely related families.

The Plastocyanin Tree

An effort recently has been made to construct a wide-spectrum phylogenetic tree based upon partial plastocyanin sequences by applying a compatability test to the sequence data set in order to recognize and correct for convergence among the sequences. Then a tree is constructed using the modified sequence set and ancestral sequence construction methods (Boulter, et al. 1979). The affinity network (unrooted tree) generated by these techniques exhibits features which remain taxonomically unsatisfying (Fig. 2).

The favorable aspect of the plastocyanin-based tree (as was also true with the tree based on cytochrome c) is that genera within accepted families cluster together (with the exceptions noted below). The seven genera of the Asteraceae, three genera of the Caprifoliaceae, four genera (five species) of the Apiaceae and two genera each of the Brassicaceae and Magnoliaceae all cluster together as groups on their respective branches. In other clusterings, however, serious problems exist. For example, two genera of the Rosaceae (*Prunus* and *Crataegus*) are separated by most of the Fabaceae; the remainder of the Fabaceae (*Phaseolus* and *Vigna*) is separated by the Brassicaceae. Additionally, the Solanaceae resides on three branches, one of which (*Capsicum*) exhibits a recent common ancestry with the Caprifoliaceae.

In order to maximize the number of taxa for which plastocyanin sequence data were available for comparison in the construction of this tree, the decision was made to determine only the 40 N-terminal residues in most cases. Such a departure in experimental approach warrants some justification.

Justification of the Use of Partial Sequences

Based upon a statistical model of evolutionary change in proteins, it can be demonstrated that the precision with which evolutionary distance between two protein sequences can be measured is inversely proportional to the square root of the number of variable residues in the sequence data set (Dayhoff 1972). Therefore, using the complete sequence of 100 residues in plastocyanin for comparisons would generate a tree 1.6 times as accurate as the tree produced using only the 40 N-terminal residues. However, since automated methods cannot be applied with ease to the determination of the remaining 60 C-terminal residue sequence, it requires 10 times as long to determine the C-terminal 60% of the plastocyanin sequence (i.e., partial sequencea can be determined 10 times as quickly as complete sequences). Since the amount of sequence data available has, in the past, been the limiting factor in tree construction, it was felt that the rapid accumulation and analysis of partial sequences more than compensated for the loss of precision relative to a tree based on completely determined sequences. It could be argued that perhaps the N-terminal 40 residues of plastocyanin were less variable than the remaining 60 and, therefore, the precision loss was greater than 1.6 by virtue of using only partial sequences. To test this possiblity among those taxa for which the complete plastocyanin sequence is known, the number of residue differences were determined among sequence pairs for the first 40 N-terminal residues and then for the 60 C-terminal residues. Among 12 sequences (66 pair comparisons), the correlation between variable residues in the N-terminal 40% and C-terminal 60% of plastocyanin was 0.72 (significant at the 0.1% level). These data indicate that the partial sequences were a representative sample of the number of variable residues within the whole plastocyanin molecule (Boulter et al. 1977).

Similarly, when separate trees were constructed first using complete sequences and then with only N-terminal partial sequences, the results were compatible within the expected limits of error (Peacock and Boulter 1975), justifying the use of partial sequences.

Additional Results with Partial Plastocyanin Sequences

Boulter, et al. (1978) have used partial sequences of plastocyanin from 22 members distributed among eight tribes of the Asteraceae (Compositae) to construct an affinity tree among those taxa. There is no consensus among taxonomists regarding tribal relationships within the Asteraceae or which tribe(s) constitute the most primitive representatives. Thus, no generally accepted "traditional" tree exists for comparison with sequence results. The tribal relationships suggested by plastocyanin sequence data are consistent with several systematic proposals made by Wagenitz (1976), and sequence data suggest that the Cynareae is the most primitive tribe of the Asteraceae. However, difficulties remain in the sequence-based tree for the Asteraceae in that several members of well-recognized tribes are separated by or intercalated (in the branching sequence) between members of very different tribes.

THE PROBLEM OF DATA HANDLING

As noted earlier, until very recently the main obstruction to constructing trees based upon amino acid sequences was the slow rate of data acquisition. With the application of automated methods and the use of partial sequences, a large amount of sequence data has been rapidly generated for analysis. During the analysis of these data, it has become increasingly clear that presently used numerical methods of data handling are inadequate, with respect to their absolute data handling capacity and their ability to deal successfully with the biological problems associated with sequence data, namely detection of an allowance for convergence among sequences. At present, the major thrust of sequence data research is in developing mathematical models and tests for convergence and multiple mutations, rather than generating additional sequences.

Methods of Tree Construction

Two basic methods of tree construction utilizing sequence data have been used most widely: the matrix method (Fitch and Margoliash 1969) and the ancestral sequence method (Dayhoff and Eck 1966). The mechanics of tree construction are detailed elsewhere (Boulter, et al. 1972) and will not be treated here. More pertinent to the present discussion is a consideration of the assumptions inherent in the application of these methods. Basic assumptions characterize both methods of tree construction.

Maximum Parsimony

Both approaches to tree construction assume that evolution has proceeded from one sequence to the next by the minimum number of mutational steps (i.e., that back, parallel, and multiple mutations at a given site do not occur or are very rare events). This assumption is clearly weak. It is estimated that parallel mutations account for 38% and back mutations for 6% of all mutational events in the cytochrome c data set (Boulter, et al. 1972 and pers. comm.). Models have been proposed based upon probability theory to correct for multiple mutations (Zuckerkandl and Pauling 1965), but these are in turn based upon the untenable assumption that amino acid changes occur randomly and with equal probability at all residue sites. It is generally accepted that a process as complex as organic evolution does not consistently proceed by the shortest, most direct path and, therefore, the assumption of maximum parsimony is a weak one.

Constant Rate of Evolution

The matrix method also assumes that the rate of evolution (i.e., fixation of amino acid changing point mutations) for a particular protein is constant over time (Jardine, et al. 1969). The validity of this assumption is dependent on the existence of a regular molecular clock. It was once thought that each protein sequence (i.e., structural gene) had a characteristic and uniform rate at which mutations were fixed over long time periods (Kimura and Ohta 1972). More recently, refined statistical tests have shown that this concept is incorrect and that neither a metronomic nor stochastic clock occurs in protein sequences (Dobzhansky, et al. 1977). The possibility of a molecular clock still exists, but only a composite clock averaged over a large number of different proteins (Fitch 1976). The assumption of a uniform rate of evolution based upon a regular clock is, therefore, a weak one.

The Global Minimum Tree is Found

The ancestral sequence method assumes that the final tree accepted as accurate is that tree (of all possible trees) which has the minimum number of amino acid substitutions (i.e., it is the global minimum tree). Since for any comparison of n species, n! possible phylogenetic trees can be drawn, this would mean that for 15 taxa there are about 10 billion possible trees!

In the construction of the cytochrome c tree only about one thousand of these possibilities could be examined. Due to the complexity of the calculations involved, the realistic absolute limit to the number of species whose sequences could be simultaneously compared is about 150 (Boulter in Cronquist 1976). This difficulty can be circumvented somewhat by constructing subtrees and consolidating them on the basis of other biological data. However, to do so compromises the objectivity of the approach and possibly violates the assumption of the global minimum tree, which could be overlooked in this construction.

From these considerations, it can be seen that the assumptions involved in tree construction are not completely satisfied in practice.

The Detection of Convergence

The data handling difficulty under most intensive investigation at present addresses the biological problem of convergence among compared sequences. Peacock and Boulter (1975) studied convergence in the cytochrome c and plastocyanin data sets and endeavored to quantify the sources of error in phylogenetic trees as a function of tree construction method and properties of the investigated protein. Through the use of computer simulation of evolutionary changes in sequences, they concluded that the ancestral sequence method was to be preferred over the matrix methods and that this method should give a reasonably accurate (2 to 8% error for cytochrome c) phylogenetic topology in spite of a fairly large number of convergent substitutions. They noted and quantitatively investigated the fact that protein properties, such as rate of evolution and proportion of convergent residue changes, affect the accuracy of a given method of tree construction.

Boulter, et al. (1979) have applied a method for detecting convergent sequence changes to a plastocyanin partial sequence data set and have constructed a network based upon the modified sequences. The convergence detection technique is based upon a character compatability test, the details of which have not been published as of this writing.

A VIEW TO THE FUTURE

At present, critically selected amino acid sequence data are insufficient and data handling methods are too flawed to produce a phylogenetic tree exhibiting a high degree of detailed accuracy. However, this approach to phylogenetic studies is a powerful one, with great potential for increasing our understanding of taxonomy and phylogeny. It is a field still in its infancy and decades will be required for its full impact to be felt. Prerequisites for that impact will include improvements in data handling methods and determination of numerous suitably chosen sequences. Even if data handling methods were perfect, so as to generate the best possible representation of relationships with a set of sequences from a single protein, it would still remain merely an affinity tree for a gene. To be confident that the sample is representative of the genome and, therefore, of the biological species, comparisons must be made among several proteins. The selected proteins must possess low convergence, compatible rates of change, and their sources must be carefully chosen so that the resultant trees are taxonomically comparable.

Current protein trees are clearly imperfect in their details, but the approach is valid. When such trees conflict with earlier schemes, they perhaps are an indication that taxonomists should carefully and critically reexamine the bases upon which their proposals have been based. It would be especially useful if taxonomists would state as explicitly as possible how their conclusions were reached. Then, with open discussions, perhaps a consensus regarding systematic relationships can be reached.

LITERATURE CITED

Behnke, H.-D. and B. L. Turner. 1971. On specific sieve-tube plastids in Caryophyllales. Further investigations with special reference to the Bataceae. Taxon 20:731-737.

Boulter, D. 1973. "The Use of Amino Acid Sequence Data in the Classification of Higher Plants". In Chemistry in Botanical Classification, edited by G. Bendz and J. Santesson, pp. 211-216. New York: Academic Press.

Boulter, D. 1974. The evolution of plant proteins with special reference to higher plant cytochromes c. Current Advances in Plant Science 8:1-16.

------. 1978. "Present Status of the Use of Amino Acid Sequence Data in Plant Phylogenetic Studies". In Evolution of Protein Molecules, edited by H. Matsubara and T. Yamanaka, pp. 243-250. Tokyo: Japan Scientific Societies Press.

Boulter, D., J. A. M. Ramshaw, E. W. Thompson, M. Richardson and R. H. Brown. 1972. A phylogeny of higher plants based on the amino acid sequences of cytochrome c and its biological implications. Proc. Royal Society (London) 181B:441-445.

Boulter, D. and J. A. M. Ramshaw. 1975. "Amino Acid Sequence Analysis of Proteins". In The Chemistry and Biochemistry of Plant Proteins, edited by J. H. Harborne and C. F. van Sumere, pp. 1-30. New York: Academic Press.

Boulter, D., B. G. Haslett, D. Peacock, J. A. M. Ramshaw and M. D. Scawen. 1977. "Chemistry, Function, and Evolution of Plastocyanin." In International Review of Biochemistry, Plant Biochemistry II, edited by D. H. Northcote, pp. 1-40. Baltimore: University Park Press.

Boulter, D., J. T. Gleaves, B. G. Haslett, D. Peacock and U. Jensen. 1978. The relationships of 8 tribes of the Compositae as suggested by plastocyanin amino acid sequence data. Phytochemistry 17:1585-1589.

Boulter, D., D. Peacock, A. Guise, J. T. Gleaves and G. Estabrook. 1979. Relationships between the partial amino acid sequences of plastocyanin from members of ten families of flowering plants. Phytochemistry 18:603-608.

Cronquist, A. 1968. The Evolution and Classification of Flowwering Plants. London: Nelson.

------. 1976. The taxonomic significance of the structure of plant proteins: A classical taxonomist's view. Brittonia 28:1-27.

Dayhoff, M. O. 1972. Atlas of Protein Sequence and Structure. Silver Spring, Md.: National Biomedical Research Foundation.

Dayhoff, M. O. and R. V. Eck. 1966. Atlas of Protein Sequence and Structure. Silver Spring, Md.: National Biomedical Research Foundation.

Dobzhansky, T., F. J. Ayala, G. L. Stebbins and J. W. Valentine. 1977. Evolution. San Francisco: W. H. Freeman.

Eckardt, T. 1976. Classical morphological features of Centrospermous families. Plant Systematics and Evolution 126:5-25.

Fitch, W. M. 1976. "Molecular Evolutionary Clocks." In Molecular Evolution, edited by F. J. Ayala, pp. 160-178. Sunderland, Massachusetts: Sinauer Associates.

Fitch, W. M. and E. Margoliash. 1967. Construction of phylogenetic trees. Science 155:279-284.

Jardine, H., C. J. van Rijsbergen and C. J. Jardine. 1969. Evolutionary rates and the inference of evolutionary tree forms. Nature, Lond. 224:185.

Kimura, M. and T. Ohta. 1972. "Population Genetics, Molecular Biometry, and Evolution." In Sixth Berkeley Symposium on Mathematical Statistics and Probability, edited by L. M. Le Cam, J. Neyman, and E. L. Scott, pp. 43-68. Berkeley: University of California Press.

Laursen, R. A., A. G. Bonner and M. J. Horn. 1975. "The Solid-Phase Peptide Sequencer". In Instrumentation in Amino Acid Sequence Analysis, edited by R. N. Perham, pp. 73-110. New York: Academic Press.

Mabry, T. J. 1973. "Is the Order Centrospermae Monophyletic?" In Chemistry in Botanical Classification, edited by G. Bendz and J. Santesson, pp. 275-285. New York: Academic Press.

Muller, J. 1970. Palynological evidence on early differentiation of angiosperms. Biological Reviews 45:417-450.

Peacock, D. and D. Boulter. 1975. Use of amino acid sequence data in phylogeny and evaluation of methods using computer simulation. Jour. Molecular Biology 95:513-527.

Ramshaw, J. A. M., D. L. Richardson, B. T. Meatyard, R. H. Brown, M. Richardson, E. W. Thompson and D. Boulter. 1972. The time of origin of the flowering plants determined by using amino acid sequence data of cytochrome c. *New Phytologist* 71:773-779.

Richardson, M., J. A. M. Ramshaw and D. Boulter. 1971. The amino acid sequence of rape (*Brassica napus L.*) cytochrome c. *Biochim. Biophys. Acta* 251:331-333.

Stevens, R. C., A. N. Glazer and E. L. Smith. 1967. The amino acid sequence of wheat germ cytochrome c. *Jour. Biological Chemistry* 242:2764-2779.

Thompson, E. W., M. Richardson and D. Boulter. 1971. The amino acid sequence of cytochrome c of *Fagopyrum esculentum* Moench (buckwheat) and *Brassica oleracea L.* (cauliflower). *Biochemistry Jour.* 124:783-785.

Wagenitz, G. 1976. Systemiatics and phylogeny of the Compositae (Asteraceae). *Plant Systematics and Evolution* 126:29-46.

Williams, J. 1974. "The Primary Structure of Proteins in Relation to Evolution." In *MTP International Review of Science, Chemistry of Macromolecules*, edited by H. Gutfreund, pp. 1-56. Baltimore: University Park Press.

Zuckerkandl, E. and L. Pauling. 1965. "Evolutionary Divergence and Convergence in Proteins." In *Evolving Genes and Proteins*, edited by V. Bryson and J. H. Vogel, pp. 97-166. New York: Academic Press.

DIVERGENCE, CONVERGENCE, AND PARALLELISM
IN PHYTOCHEMICAL CHARACTERS: THE
GLUCOSINOLATE-MYROSINASE SYSTEM

James Eric Rodman

Department of Biology
Yale University
New Haven, Connecticut   06520

The very first human response to glucosinolates (mustard oil glucosides) may well have been a gastro-intestinal one: specifically, a belch. The hydrolytic products or mustard oils of radishes and cabbages, for example, can induce that reaction. Scientific investigations have added a new perspective to mankind's long-standing appreciation of these compounds in their role as flavorings and condiments (Vaughan and Hemingway 1959), and to a perhaps equally long-standing antipathy to them, in their roles as allergens and irritants (Mitchell and Jordan 1974; VanEtten and Tookey 1979). While these descriptive and experimental studies over the last century and a half have made this class of compounds one of the best-known groups of secondary plant metabolites, new problems have been raised--biosynthetic, developmental, ecologic, and phylogenetic--about the place of glucosinolates in the economy of nature. The glucosinolate-synthesizing taxa have presumably evolved from non-glucosinolate lineages, and the fundamental taxonomic question is whether this divergence was a single event or whether it occurred more than once. That is to say, are the glucosinolate-producing taxa a monophyletic assemblage, or are the glucosinolate-myrosinase systems polyphyletic in origin and hence not strictly homologous? Related to this taxonomic question are the fundamental evolutionary problems of why and how this thioglucoside-thioglucosidase system originated and diversified. The system, therefore, presents problems which can be framed under the broad topics of divergence, convergence, and parallelism (see Davis and Heywood 1963: 34ff, for a lucid discussion of the terms and concepts employed here). The attack on these problems can serve as a model for the evaluation and taxonomic exploitation of any group of phytochemical characters. This review will concentrate on the taxonomic and evolutionary problems, and will map out promising new roads for future systematic studies. The fine reviews by Kjaer (1960) and Ettlinger and Kjaer (1968) provide a comprehensive introduction to the chemistry and biology of glucosinolates; several shorter reviews by Kjaer (1963, 1966, 1974, 1976, 1977) supplement these. Tapper and Reay (1973) and VanEtten and Tookey (1978, 1979) have reviewed the glucosinolate-myrosinase system fron an agronomic perspective.

THE GLUCOSINOLATE-MYROSINE SYSTEM: SOME CHEMISTRY, ANATOMY, AND GENETICS

Glucosinolates are a class of thioglucosides, that is, sulfur-linked glucosides (Fig. 1), with variations thus far reported only in R, the side-chain of the molecule. Josefsson (1970)

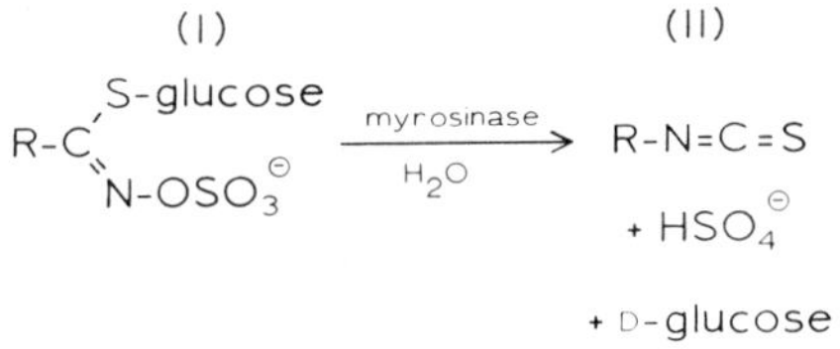

$$\underset{\substack{(\text{I}) \\ R-C\diagdown_{N-OSO_3^\ominus}^{S\text{-glucose}}}}{} \xrightarrow[H_2O]{myrosinase} \underset{\substack{(\text{II}) \\ R-N=C=S \\ + HSO_4^\ominus \\ + \text{D-glucose}}}{}$$

Figure 1. General reaction for the enzymatic hydrolysis of glucosinolates (I) to their corresponding isothiocyanates (II).

estimated that fully half the sulfur in seeds of *Brassica* crop plants is bound in glucosinolates. The compounds thus constitute a significant item in the sulfur and mineral budget of these plants. Unfortunately, little work has been done on ontogenetic changes in glucosinolate composition and concentration; we do not know, for example, if the large amount of sulfur in seed glucosinolate serves as a reservoir in seedling protein synthesis. Indeed, the catabolism of glucosinolates in plants is virtually unknown (Bergmann 1970; Sørensen 1970; Elliott and Stowe 1971b).

Approximately 85 different glucosinolates are now known structurally, with side-chains varying from a simply methyl group to more complex substituents. Ettlinger and Kjaer (1968) proposed a useful classification of side-chain chemical types, based on structure and on likely biosynthetic derivation. Most new glucosinolates discovered since their review can conveniently be placed on this scheme (see Kjaer 1974 for a recent listing of types). Many glucosinolates are formally analogous to amino acids, and several have been shown to be biosynthesized from protein amino acids or from those that have undergone carbon chain extension or other modifications (see reviews by Kjaer and Larsen 1973, 1976; Tapper and Reay 1973; Underhill et al. 1973). Biosynthesis from precursor amino acids, via compounds called aldoximes, appears to hold for most if not all glucosinolates, although it should be stressed that a few proposed mustard oil glucosides would require unlikely amino acid precursors (e.g., 3,4-dimethoxyphenylglucosinolate, proposed by Danielak and Borkowski 1970; cf. Underhill et al. 1973).

Glucosinolates are usually present throughout the plant, and are thought to be localized in cell vacuoles. Grob and Matile (1979) presented evidence from cell-fractionation studies for the sequestration of glucosinolates in vacuoles of root cells of horseradish, *Armoracia lapathifolia* Gilib. (Cruciferae). However, glucosinolates are also richly present in mature, dormant seeds, presumably not in aqueous vacuoles, so the problem of cellular localization remains open.

Without known exception, glucosinolates are accompanied in the plant by a specific enzyme, classically known as myrosin and less classically as myrosinase, glucosinolase, sinigrinase, or thioglucosidase (thioglucoside glucohydrolase, E.C. 3.2.3.1; cf. Tapper and Reay 1973). When brought into contact with glucosinolates, the enzyme catalyzes the hydrolysis of the thio-glucoside bond and leads to the formation, typically, of mustard oils or isothiocyanates (Fig. 1; Ettlinger and Lundeen 1956). Certain pH conditions and possibly accessory enzymes, which have thus far eluded characterization (Benn 1977; Luthy and Benn 1977), promote the formation of alternative hydrolysis products such as thiocyanates or nitriles (VanEtten and Tookey 1979). All of these are physiologically potent compounds for a variety of organisms (VanEtten and Tookey 1979). Myrosinase is now known to be a family of isoenzymes, and recent work on *Brassica* and *Sinapis* species (Cruciferae) suggests that different enzyme morphs may be taxonomically diagnostic (see Bjorkman 1976 for a review). Comparison and characterization of the myrosinase isozymes from all the glucosinolate-producing plant families, by electrophoresis or preferably by amino acid sequencing, will be especially helpful in resolving the question of homology of the glucosinolate-myrosinase systems.

The myrosinase enzyme is thought to be concentrated in specialized protein-rich idioblasts, called myrosin cells, which stain bright red with Millon's reagent (Fahn 1979: 147ff). Other stains can also be employed to visualize myrosin cells, for example, orcein solution and concentrated hydrochloric acid, lactophenol aniline blue (Fahn 1979), and even the standard safranin-fast green (Werker and Vaughan 1974, 1976). It must be emphasized, however, that none of these staining techniques, including the use of Millon's reagent, is uniquely specific for myrosinase; rather, the stains react with a class of proteinaceous (and possibly other organic) material (Johansen 1940: 185; Baker 1956). Iversen (1970) and Pihakaski and Iversen (1976) have employed histochemical tests, in conjunction with electron microscopy, that make use of specific products of myrosinase hydrolysis to localize the enzyme. They found myrosinase activity in association with the endoplasmic reticulum of cells throughout the root tip of *Sinapis alba* L. and not in specialized myrosin cells. Jørgensen et al. (1977), however, demonstrated that Millon-staining myrosin cells contain material that reacts positively when histochemically tested for protein and for glycols, and it is known that myrosinase isoenzymes are glycoproteins (Björkman 1976). What precisely myrosin cells are remains a problem. It is possible that the myrosinase enzyme is present in low concentrations throughout most or all cells of the plant, where it may function in the metabolic turnover of glucosinolates, and that furthermore it is sequestered in large amounts in scattered myrosin idioblasts, where it is available for rapid hydrolysis of the thioglucosides upon disruption or injury of the tissue.

Since Heinricher's (1884, 1886) pioneering work on these specialized cells, anatomical studies have documented differences in the morphology and distribution of myrosin cells in glucosinolate-producing plants (e.g., Guignard 1890, 1893; Jørgensen et al. 1977). Indeed, von Hayek (1911) emphasized the distribution of myrosin cells in leaves as a generic and tribal character in his classification of the Cruciferae. Fahn (1979), in his review of plant secretory tissues, wrote that myrosin cells are absent from Euphorbiaceae, Plantaginaceae, and Gyrostemonaceae (listed under Phytolaccaceae by Fahn), members of which have been reported to contain glucosinolates or their hydrolysis products. In fact, Millon-stained myrosin cells have been reported to occur in *Drypetes (=Putranjiva) roxburghii* (Wall.) Hurusawa, a euphorbiaceous species known to produce glucosinolates (Jorgensen et al. 1977). For Plantaginaceae, neither myrosin cells nor glucosinolates have been conclusively documented (see below). Carlquist (1978) reported that Bataceae and Gyrostemonaceae lack myrosin cells, but he provided no documentation or description of methods used to test for these. Behnke (1977) found protein deposits in phloem companion cells of Gyrostemonaceae, which he speculated might be related to similar inclusions associated with dilated cisternae of cells of Capparales, which in turn might be sites of myrosinase deposition (Behnke 1977). Apparently none of the Caricaceae, Salvadoraceae, or Tovariaceae (cf. Cronquist 1968:213) has been analyzed critically for myrosin cells, using methods such as those described by Jorgensen et al. (1977). In other families like Capparaceae, Cruciferae, Resedaceae, Limnanthaceae, and Tropaeolaceae, and in *Drypetes* of Euphorbiaceae, the glucosinolate-myrosinase system constitutes a coordinately evolved biochemical-anatomical phenomenon. Clearly, however, extensive and careful systematic surveys are needed, both to substantiate the suggested taxonomic differences in myrosin cell morphology and distribution, and to determine the chemistry and ontogeny of these unique cells. In addition, future anatomical and histochemical studies should focus on genera that have, in the hands of various phylogenists, been related to glucosinolate-producing taxa; for example, such problematic genera as *Akania, Bretschneidera* (reported to contain myrosin cells by Tang, cited in Carlquist 1978, but not known to produce glucosinolates), *Emblingia, Koeberlinia* (reported not to contain glucosinolates: Ettlinger and Kjaer 1968; Rodman unpublished results for fruits and seeds), *Oceanopapaver, Pentadiplandra* (recently reported to produce a novel isothiocyanate, but from an as yet uninvestigated precursor: El Migirab et al. 1977), and *Physena*.

It is presumably only upon some kind of mechanical disruption or injury to the plant that the myrosinase enzyme and a glucosinolate substrate are brought together, and one then observes the production of isothiocyanates or other hydrolysis products. This cellularly comparmentalized thioglucoside-thioglucosidase system is analogous to compartmentalized systems in other plants, such as the widespread cyanogenic glycoside-glycosidase system (Conn 1979). There is, additionally, some evidence for the endogenous production of isothiocyanates by intact plants or plant parts, at very low levels (Tang 1971; Mahadevan 1973; see Harborne 1977a: 59). If confirmed as a general phenomenon, it can be concluded that glucosinolate-producing plants would emit a characteristic chemical aroma or "halo" (Hanover 1975) comprising small amounts of volatiles released by the hydrolysis of glucosinolates. Virtually all the hydrolysis products are physiologically potent compounds, indeed, destructive in sufficient quantity, for microorganisms, plants, and animals (Whittaker and Feeny 1971; Feeny 1977; VanEtten and Tookey 1979). This point will be raised again in the consideration of function of these secondary plant metabolites.

Formal study of the genetics of glucosinolate composition has largely concerned *Brassica* crop plants, because of their commercial importance and specifically because of an interest in breeding oilseed cultivars low in toxic glucosinolates (VanEtten et al. 1969; Josefsson 1970). These studies have confirmed the genetic control over glucosinolate composition (that is, kinds of constituents present) as well as control of total quantity of thioglucosides, although quantity has been found to be subject to environmental influence (e.g., by mineral nutrition; see Josefsson 1970, for review). Plants producing several glucosinolates manifest polygenic control, and the segregation patterns may be complex, at least in the Brassicas investigated (Kondra and Stefansson 1970). Working on wild taxa of *Cakile* (Cruciferae), Rodman (1974) demonstrated that artificial F_1 hybrids were qualitatively additive (for kinds of glucosinolates) and quantitatively intermediate (for relative proportions) in their glucosinolate profiles. The F_2 generations from these artificial crosses in *Cakile* showed a broad spectrum of glucosinolate arrays, indicating segregation of the genes controlling for kinds and amounts of thioglucosides (Rodman 1978b, and in preparation). Nothing, unfortunately, is known about the genetics of the myrosinase enzymes. Since glucosinolates seem always to be accompanied in plants by myrosinase (Ettlinger and Kjaer 1968), it would be interesting to know if genes controlling for the two products are linked on the same chromosome.

TAXONOMIC DISTRIBUTION AND DIVERSITY

The taxonomic distribution of glucosinolates and myrosinase is now fairly well known, due largely to the work of Martin Ettlinger, Rolf Gmelin, Anders Kjaer, and their colleagues. Surprises, if any, will likely come from tropical plant families. The Glucosinolate-myrosinase system is restricted to dicotyledonous angiosperms (Magnoliatae or Magnoliopsida). Its principal locus is the Cruciferae-Capparaceae alliance, in terms both of number of plant species and of diversity of glucosinolate structures (side-chains). Furthermore, every species of Cruciferae tested thus far and most Capparaceae as well have been found positive for glucosinolates, the number of compounds identified from an individual plant varying from one to 15 (Rodman 1974; Van Etten et al. 1976). In total, about 12 families are known to produce glucosinolates (and presumably myrosinases) (Table I). In all of these except Euphorbiaceae, glucosinolates are thought to be present in most if not all species. In the large and diverse Euphorbiaceae, only *Drypetes (=Putranjiva) roxburghii* (Kjaer and Friis 1962) and *D. gossweileri* S. Moore (Ettlinger and Kjaer 1968) are known to produce glucosinolates; *Guya (=Drypetes) caustica* Frapp. ex Corden. may also contain these compounds (M. G. Ettlinger, personal communication). Many other Euphorbiaceae were found to lack glucosinolates and myrosinases (Ettlinger and Kjaer 1968). For the small family Plantaginaceae, Prochazká (1959) reported 4-methylsulfinyl-3-butenyl isothiocyanate from leaf hydrolysates of a *Plantago* species, a report which was questioned by Ettlinger and Kjaer (1968). Recently, Cole (1976) reported isopropyl isothiocyanate and 5-vinyl-2-oxazolidinethione as autolysis products from leaves of *Plantago major* L. Bona fide glucosinolates have not yet been documented for the family, and a concerted program of isolation and structural determination of the parent thioglucosides, if present at all, is required to resolve the issue.

Glucosinolate diversity varies widely in this group of families. On the basis of current data (and exluding the atypical Euphorbiaceae from the calculation) there is an interesting positive correlation ($r = 0.99$) between species diversity within a family and structural variation in glucosinolates. This suggests that glucosinolate diversification has accompanied speciation in these families. The limited diversity in certain small families may or may not be real. The Caricaceae, for example, have been investigated repeatedly, and only benzylglucosinolate has been reported from these plants (Ettlinger and Hodgkins 1956; Gmelin and Kjaer 1970; Tang 1971; Tang et al. 1972; Tang and Hamilton 1976; Flath and Forrey 1977; Chan et al. 1978; Rodman, unpublished data on seeds), although Schraudolf (1965) tentatively identified an indolic glucosinolate in *Carica papaya* L. seedlings. In other instances, however, the limited diversity of glucosinolate structures may be an artifact of limited sampling of species or of techniques with poor resolving power (e.g., paper chromatography in a single solvent system). The Tropaeolaceae provide an instructive example. Until recently,

Table 1. Taxonomic distribution and diversity of glucosinolates. Estimated number of species per family taken from Airy Sahw (1966). Number of species reported positive for glucosinolates (or for their hydrolysis products) and number of glucosinolates identified per family compiled or estimated from the literature.

Family	Estimated Number of Species in Family	Number of Species Reported Positive for Glucosinolates	Number of Glucosinolates Identified for Family
Cruciferae	3200	400	75
Capparaceae	925	50	20
Resedaceae	70	8	9
Moringaceae	12	2	6
Tovariaceae	2	1	3
Caricaceae	55	15	1
Euphorbiaceae	5000	2	4
Salvadoraceae	12	1	1
Limnanthaceae	11	7	2
Tropaeolaceae	92	10	9
Bataceae	2	1	1
Gyrostemonaceae	16	2	3

only the widespread benzyl, isopropyl, and 2-butyl glucosinolates were knwon from this family. Using gas chromatography and mass spectrometry, Kjaer et al. (1978) found nine different glucosinolates in various species of *Tropaeolum*. Similarly for the Moringaceae, Kjaer et al. (1979) have increased from two to six the number of identified compounds in this small family. In those families where chemical diversity is low, the characteristic glucosinolates are usually the benzyl, isopropyl, 2-methylpropyl, or 2-butyl compounds; these are also found in the larger families. The compounds are apparently biosynthesized from the protein amino acids phenylalanine, valine, leucine, and isoleucine, respectively (Kjaer and Larsen 1973, 1976). Kjaer and Malver (1979) have suggested that biosynthesis from protein amino acids is a primitive phylogenetic trait whereas chain extension (homologization) or other modifications of the precursor amino acids constitute advanced evolutionary features (cf. also Kjaer et al. 1979). Their views are strictly in line with the widespread assumption among phylogenists that "common is primitive" (Estabrook 1977), and furthermore they reflect the assumption that biosynthetic pathways leading simply and directly from common precursors are likely to be more primitive phylogenetically than pathways which are more elaborate or complex. This again represents the widespread assumption among phylogenists that early ontogenetic stages of development (in a chemical context, initial biosynthetic steps) are also phylogenetically primitive (see Gould 1977, for a review).

Glucosinolate diversity is not random. Taxon-specific glucosinolates may distinguish among families of Table 1. Methylglucosinolate, for example, is common in plants of the Capparaceae but appears to be totally absent from the closely related Cruciferae (Kjaer 1974). Rhamnose-substituted side-chains may prove to be common in Moringaceae (Kjaer et al. 1979) and Resedaceae (Olsen and Sørensen 1979) but absent or rare in other families. Within chemically diverse families such as Cruciferae, and most likely Capparaceae (Ahmed et al. 1972) as well as Tropaeolaceae (Kjaer et al. 1978), glucosinolate profiles (kinds and proportions of constituents) are taxon-specific, although specificity may vary over the levels of the taxonomic hierarchy within a family and can only be established empirically. Kjaer and Hansen (1958) first demonstrated in a paper-chromatographic study of some *Arabis* species that particular arrays of glucosinolates may distinguish species. Subsequent extensive sampling, usually with paper and/or gas chromatography, has documented cultivar- or species-specific glucosinolate profiles within the genera *Brassica* (Ettlinger and Thompson 1962; Phelan and Vaughan

Table 2. Seed glucosinolate profiles of the 14 taxa of *Cakile* (Cruciferae), determined by gas chromatography of volatile isothiocyanates for first 12 compounds and by paper chromatography of thiourea derivatives for last 5 compounds. Average percentage peak-area and standard deviation computed by arcsine transformation (Sokal and Rohlf 1969: 386ff) for all volatile compounds constituting, in at least one sample of that taxon, more than 10% peak-area on a gas-chromatogram obtained under

	C. arabica (2 + 1*)	*C. maritima* ssp. *euxina* (1 + 1)	*C. maritima* ssp. *maritima* (western Mediterranean; 4 + 2)	*C. maritima* ssp. *maritima* (Western Europe; 6 + 2)	*C. maritima* ssp. *baltica* (8 + 5)	*C. artica* (1 + 0)	*C. edentula* ssp. *edentula* var. *lacustris* (4 + 0)	*C. edentula* ssp. *edentula* var. *edentula* (northern race; 6 + 2)
Unknown 1	0	0	0	0	0	0	0	0
$(CH_3)_2CH$	59(0.62)	0	24(0.42)	<5	T	0	T	<5
$CH_2=CHCH_2$	23(0.70)	99	5(6.87)	11(1.07)	84(2.84)	92	94(0.01)	56(4.91)
$C_2H_5CH(CH_3)$	17(0.04)	0	34(1.99)	86(1.32)	14(3.02)	0	<5	40(6.27)
$CH_2=CH(CH_2)_2$	T	0	<8	0	T	T	<2	<2
$CH_2=CH(CH_2)_3$	0	0	0	0	0	0	0	0
$CH_3S(CH_2)_3$	0	T	3(2.08)	T	T	6	T	<2
$C_6H_5CH_2$	0	0	0	0	0	0	T	T
$CH_3S(CH_2)_4$	0	0	9(4.18)	0	0	T	0	0
$C_6H_5(C\ _2)_2$	0	0	0	0	0	T	0	0
$CH_3S(CH_2)_5$	0	0	8(4.39)	0	0	0	0	0
$CH_3S(CH_2)_6$	0	0	T	0	0	0	0	0
$CH_3SO(CH_2)_3$	0	0	0	0	0			?
$CH_3SO(CH_2)_4$	0	0	+	0	0			0
$CH_3SO(CH_2)_5$	0	0	+	0	0			0
$CH_3SO_2(CH_2)_4$	0	0	0	0	0			0
$CH_3SO_2(CH_2)_5$	0	0	+	0	0			0

uniform operating conditions (peak-areas measured by trangulation); * = number of gas-chromatographic samples plus number of paper-chromatographic samples analyzed. Unknown 1 tentatively identified as 3-butenyl cyanide, an alternative hydrolysis product from 3-butenylglucosinolate. T = trace amount; + = present; blank = not determined. Compiled from published analyses in (Rodman 1974, 1976).

C. edentula ssp. edentula var. edentula (southern race; 14 + 11)	C. edentula ssp. harperi (10 + 3)	C. constricta (4 + 1)	C. geniculata (8 + 2)	C. lanceolata ssp. lanceolata (11 + 2)	C. lanceolata ssp. fusiformis (9 + 2)	C. lanceolata ssp. pseudoconstricta (3 + 1)	C. lanceolata ssp. alacranensis (1 + 0)
0	8(0.19)	<10	<9	6(0.35)	5(0.73)	6(1.06)	0
T	0	0	0	0	0	0	0
91(0.43)	<5	9(0.26)	17(1.06)	6(1.75)	8(0.13)	10(2.96)	0
6(0.42)	T	<8	T	T	T	T	0
<3	85(0.39)	76(1.09)	73(0.90)	80(1.66)	79(0.95)	78(3.97)	53
0	<5	0	0	T	T	T	0
<4	T	T	<3	<3	<3	<3	0
T	0	0	0	0	0	0	0
0	<3	<5	<4	5(1.20)	<8	<4	47
0	T	0	0	T	0	0	0
0	T	T	0	T	T	T	0
0	0	0	0	0	0	0	0
?	0	0	0	0	0	0	
0	+	?	0	0	0	0	
0	0	0	0	0	0	0	
0	?	0	0	0	0	0	
0	0	0	0	0	0	0	

1976; VanEtten et al. 1976; Daxenbichler et al. 1979), *Cakile* (Rodman 1974, 1976) and *Thelypodium* (Al-Shehbaz 1973) in the Cruciferae. Many, if not most, species of the large genus *Streptanthus* (Cruciferae) appear distinguishable on the basis of seed glucosinolate composition (Rodman and Kruckeberg, in preparation). Numerous other literature reports, on single or few species, in the aggregate suggest great potential chemotaxonomic utility for glucosinolates. In all cases, however, extensive sampling is required since it is clear that species may be chemically polymorphic. Geographical variation in glucosinolate composition has been documented, for example, in cultivated *Brassica juncea* Coss. (Vaughan and Gordon 1973), in weedy *Lepidium bonariense* (Kjaer et al. 1971), and in wild *Cakile edentula (Bigelow)* Hook. and *C. maritima* Scop. (Rodman 1974, 1976). Whether closely related species or genera differ or not in glucosinolate composition can only be settled empirically since there is yet no a priori reason for expecting particular patterns of variability in these compounds.

For assessing generic and tribal relationships within the large familiy Cruciferae and the smaller Capparaceae, the data are sparse, but the evidence suggests a useful role for glucosinolate characters, contrary to the early pessimism of Daxenbichler et al. (1964). The situation will also improve as more taxa are studied with sensitive chromatographic techniques and as taxa are characterized not simply by presence/absence of particular glucosinolates but on the basis of biosynthetic pathways (Rodman 1976; Kjaer et al. 1978; Kjaer et al. 1979). At the moment, structurally identified glucosinolates have been reported in the literature for about 70-80 genera of Cruciferae. For comparison, chromosome counts have been published for over 150 genera of crucifers, providing nearly 50% generic coverage for the family (Rodman, compilation). Like the chromosome data, most of the glucosinolate reports are based on one or a few samples, while very few genera have been surveyed intensively and systematically. Nonetheless, suprageneric patterns are emerging from this growing body of chemical data. Only a few examples will be mentioned here. Schulz's (1936) tribe Brassiceae, generally considered a natural phyletic assemblage (Hedge 1976), is characterized primarily by methionin-derived allyl and 3-butenyl glucosinolates and their biosynthetically related methylthioalkyl and methylsulfinylalkyl analogues; these compounds have been reported in at least 15 genera in five of Schulz's seven subtribes (the Vellinae and Savignyinae not yet having been investigated; Rodman, compilation). In contrast, Schulz's tribe Euclidieae has been recognized as a clearly unnatural grouping (Hedge 1976). The heterogenous glucosinolate data, although based on limited sampling, reinforce this judgement, without necessarily indicating the true relationships of these genera: *Boreava* (Daxenbichler et al. 1964) and *Schimpera* (Rodman unpublished data) produce 3-butenylglucosinolate; *Bunias*, isopropylglucosinolate (Cole 1976); *Ochthodium* isopropyl and 2-butyl glucosinolates and their hydroxy analogues (Rodman unpublished data); and *Neslia*, the 9-, 10-, and 11-methylsulfinyl-alkyl compounds (Kjaer and Schuster 1972a). Other examples can be cited for the Cruciferae where morphologically related clusters of genera share biosynthetically related groups of glucosinolates: *Arabis-Sibara-Nasturtium-Pennellia* (MacLeod and Islam 1975; Rodman unpublished data); *Cakile-Erucaria* (Rodman 1976 and unpublished data); *Caulanthus-Streptanthus-Thelypodium-Warea* (Al-Shehbaz 1973; Rodman unpublished data); *Erysimum-Malcolmia-Syrenia* (Kjaer and Schuster 1972b). As data accumulate, these and other clusters may serve as core groupings on which a natural tribal framework for the ca. 350 genera of Cruciferae can be constructed. Comparable studies of the chemically diverse Capparaceae (see Rodman 1978 for references) should also prove useful in the construction of an infrafamilial classification of these plants.

In studies at infrageneric levels, glucosinolate composition has been demonstrated to provide valuable taxonomic characters in the genera *Brassica, Cakile, Streptanthus,* and *Thelypodium* (see references above). From 15 to 25 different glucosinolates were identified in these various genera, and different subsets of the total, differing in kinds and/or proportions, were found to correlate with morphologically recognized subspecies, species, or species-groups. Table 2 summarizes the published seed glucosinolate profiles of *Cakile* taxa as an example (Rodman 1974, 1976). Comparative studies must, of course, be organ-specific since it is well established that glucosinolate composition and concentration may vary among plant parts (Josefsson 1967) and also vary with developmental age (Delaveau 1952, 1958; Cole 1978). Profiles of glucosinolates characteristic of subspecies, species, or species-groups have also been documented for *Capparis* (Capparaceae; Ahmed et al. 1972) and for *Tropaeolum* (Tropaeolaceae; Kjaer et al. 1978), although on the basis of less extensive sampling. Glucosinolate analyses can be especially rewarding in biosystematic studies within a family because the compounds provide a clearly defined set of genetically determined, homologous characters, which can be surveyed efficiently in numerous samples with appropriate techniques (Rodman 1978a, b).

POPULATION BIOLOGY AND ECOLOGY OF GLUCOSINOLATES

On the population level, the biology of glucosinolates has been studied both comparatively and experimentally from the complementary perspectives of biosystematics and ecology. Indeed, the population biology of glucosinolates is probably the best studied of that of any comparably diverse class of secondary plant metabolites, due to a long-standing fascination with the ecological roles of these compounds (e.g., Verschaffelt 1911; Whittaker and Feeny 1971; Chew and Rodman 1979), to more recent biosystematic interests, and of course to an outstanding tradition of chemical research by such individuals as Nobelist A. I. Virtanen (e.g., Virtanen 1965), to name only one of the more illustrious. It is sobering to realize, therefore, that we remain largely ignorant of the patterns of variability in glucosinolates within natural plant populations, of the physiological and ecological roles of these compounds, and of the processes of evolutionary change in thioglucoside composition.

Within natural populations, glucosinolate composition may be remarkably homogeneous. Table 3 documents the minor intrapopulational variation detected in seed isothiocyanates of *Cakile edentula* ssp. *edentula* and ssp. *harperi* (Small) Rodman and in leaf isothiocyanates of *Descurainia richardsonii* (Sweet) O. E. Schulz. Recently, however, Rodman and Chew (1979) have reported much greater variation in leaf isothiocyanate profiles in two Colorado populations of *Arabis drummondii* Gray and *Cardamine cordifolia* Gray (Cruciferae). Table 4 records the leaf profiles detected in 25 individuals from the population of *A. drummondii*. Polymorphism in these two cases has not yet been proven to be genetic (Rodman and Chew 1979). Since geographical variation in glucosinolate composition has been shown for several species (references cited above), other instances of intrapopulational polymorphism are likely. Currently, however, data from adequate intra- and interpopulational surveys are too sparse to determine if there are patterns in glucosinolate composition correlated with life history (annual vs. perennial) or with physical or biotic variables of the habitat (e.g., moisture, light, mineral nutrients; numbers of competing plants, numbers of herbivores or pathogens).

The known patterns of glucosinolate variability do not appear to be compatible with the several physiological roles proposed for these compounds (see Rodman 1978a, for a listing with

Table 3. Intrapopulational variation in glucosinolate composition, determined by gas chromatography of volatile isothiocyanates, for seeds of *Cakile edentula* ssp. *edentula* var. *edentula* (one population) and *C. edentula* ssp. *harperi* (two populations) and for leaves of *Descurainia richardsonii* (one population). Average percentage peak-area and standard deviation computed by arcsine tranformation (Sokal and Rohlf 1969:386ff) for all compounds constituting, in at least one individual of that population, more than 5% peak-area on a gas-chromatogram obtained under uniform operating conditions (peak-areas measured by triangulation). Unknown 1 tentatively identified as 3-butenyl cyanide, an alternative hydrolysis product from 3-butenylglucosinolate; unknown α tentatively identified as allyl thiocyanate, an alternative hydrolysis product from allylglucosinolate. T=trace amounts. Compiled from unpublished seed data (cf. Rodman 1978b) and from Rodman and Chew (1979).

Compounds	*Cakile edentula*[a] ssp. *edentula* var. *edentula* (so. race)	*C. edentula* ssp. *harperi* 1[b]	2[c]	*Descurainia richardsonii*[d]
Unknown isopropylglucosinolate	0	13(0.41)	12(0.29)	
isopropylglucosinolate	T	0	0	T
allyl-	87(0.09)	3(0.03)	3(0.07)	2(0.47)
Unknown α	7(0.01)	0	0	
2-butyl-	4(0.26)	T	T	T
3-butenyl-	2	83(0.24)	83(0.21)	97(0.37)
4-pentenyl-	0	T	T	T
3-methylthiopropyl-	T	0	0	
benzyl-	T	0	0	T
4-methylthiobutyl-	0	T	T	
2-phenylethyl-	0	T	T	
5-methylthiopentyl-	0	T	T	
isobutyl				T

[a] North Carolina, Dare Co., Southern Shores ocean beach north of Kitty Hawk (R371, YU; N=10)

[b] North Carolina, Carteret Co., ocean strand east of Emerald Isle (R365, YU; N=12)

[c] North Carolina, New Hanover Co., Carolina Beach ocean strand near public fishing pier (R366, YU; N=12)

[d] Colorado, Gunnison Co., montane meadows along Copper Creek ca. three miles above Gothic (N=14)

Table 4. Leaf glucosinolate composition of 25 individual and 6 bulk samples of *Arabis drummondii* (Cruciferae) from Copper Creek population (Colorado, Gunnison Co., montane meadows along Copper Creek ca. three miles above Gothic), determined by gas chromatography of volatile isothiocyanates for individual samples (percentage peak-area measured by triangulation) and by paper chromatography of hydrolysis products for bulk samples. The 25 individual samples of collection R402 further tested for cyclic oxazolidinethiones by paper chromatography of extracts originally prepared for GC (cf. Rodman 1978a). T=trace amount (less than 1% of total peak area); ++ = major componuent; + = minor or trace compouent; * = sample split to test for hydrolysis products from endogenous leaf myrosinase vs. exogenous (*Sinapis alba*) myrosinase activity, with no differences detected. Note reversal in relative amounts of isopropyl and 2-butyl glucosinolates in individuals R402-1 and R402-2; and occurrence of the hydroxy analogue of isopropylglucosinolate as major leaf compound in individual R402-13. From Rodman and Chew (1979).

COMPOUND

Collection number	isopropyl-glucosinolate	2-hydroxy-1-methlethyl-	2-butyl-	5-methyl-sulfinylpentyl-	6-methyl-thiohexyl-	6-methyl-sulfinylhexyl-	6-methyl-sulfonylhexyl-	benzyl-
	(percentage peak-area from gas-chromatogram)							
R402-1	99		1					t
R402-2	98		2					t
R402-3	4		96					
R402-4	3		97					t
R402-5	5		95					t
R402-6	5		95					t
R402-7	4		96					t
R402-8	5		92	2				t
R402-9	5		92	2				1
R402-10	4		96					t
R402-11	3		94	2				t
R402-12	5		92	3				
R402-13		++						t
R402-14	3		97					t
R402-15	4		96					
R402-16	4		96					
R402-17	3	+	97					t
R402-18	5		95					
R402-19	4		96					t
R402-20	4		96					
R402-21	5		93	2				t
R402-22	4		96					t
R402-23	5		95					
R402-24	3		97					
R402-25	2		98	t				
	(relative amounts determined by paper chromatography)							
R267	+		++		+	++	+	
Chew s.n. 1974	+		++		+	+		
C 2	+	+	++				+	
C 21			+			++		
C31			+			++		
C 36/37*	+	+	++		+	+	+	

references). As suggested waste products, for example, glucosinolates are incongruous in not being accumulated in senescent tissues but rather being concentrated in inflorescences and seeds (Tapper and Reay 1973; VanEtten and Tookey 1979). As suggested metabolic regulators, the compounds appear strange because they are variable among taxa and occasionally polymorphic within populations. The proposed role of indolic glucosinolates as auxin precursors, analogues, or sinks (e.g., Gmelin and Virtanen 1961; Andersen and Muir 1966; Kutáček and Kefeli 1968) has received little support (e.g., Elliott and Stowe 1971a) and rather more opposition (e.g., Bergmann 1968, cited in Josefsson 1970; Schraudolf and Weber 1969; Josefsson 1970). A role for glucosinolates in the metabolism of amines represents perhaps the most promising line of current physiological research (e.g., Sørensen 1970; Dalgaard et al. 1977; Olsen and Sørensen 1979). Of course, physiological roles are not mutually exclusive with other postulated functions, and the disquieting fact remains that little work has been done on the ontogeny and catabolism of glucosinolates.

The alternative, ecological hypothesis of chemical defense by glucosinolates has been argued cogently and frequently (e.g., Fraenkel 1959; Whittaker and Feeny 1971; Chew and Rodman 1979), and moreover, the ecological hypothesis has proved to be excitingly heuristic. Much of the circumstantial and experimental evidence for an allelochemic role for glucosinolates and their hydrolysis products is summarized by Feeny (1977), Chew and Rodman (1979), and Rodman and Chew (1979). While insect herbivores have been the focus of much of this literature, fungal pathogens may prove to be more important objects of glucosinolate-mediated plant defenses (see Walker et al. 1937; Pryor et al. 1940; Lewis and Papavizas 1971; Greenhalgh and Mitchell 1976). The evolutionary mechanism of "stepwise reciprocal selective response" between chemically defended plants and their herbivores and/or pathogens, first outlined in detail by Ehrlich and Raven (1965), provides a plausible rationale for the observed variation in glucosinolate composition among taxa and within populations. The presence of a chemical defense in plants selects for counter-adaptations in herbivore or pathogen populations, and successful heterotroph adaptation in turn selects for divergent plant defenses, either for diversification of existing biochemical defenses (Rodman and Chew, 1979) or for evolution of novel chemical, physical, or biotic defense mechanisms. Evolutionary shifts by plants in their allelochemic defenses, in response to selection operating through differential herbivore and/or pathogen attack, would result in the divergent glucosinolate profiles which have been found to characterize related taxa. Conceivably there could also be convergence on a particular glucosinolate array by different taxa, where such a profile conferred resistance to herbivores or pathogens, but diversification appears to be the more likely route within any one community. Recently published data (Rodman and Chew 1979) on chemical diversity in a natural community of several glucosinolate-producing species are consonant with the prediction that antagonistic plant-heterotroph interactions should lead to evolutionary diversification of plant defenses (Janzen 1973; Feeny 1976; Levin 1976; Cates and Rhoades 1977). It is thus to be expected that different glucosinolates should exert different effects on herbivores or pathogens, and this has been documented in several laboratory experiments (see references in Rodman and Chew 1979). For natural communities, correlative evidence now suggests that different glucosinolates mediate differential responses of a specialist insect herbivore to its potential cruciferous food plants (Rodman and Chew 1979; but see Nielsen et al. 1979, for a different viewpoint based on studies with different insects). Experimental proof of the differential effects of glucosinolates in an ecologically relevant context must come from studies on chemically polymorphic plant populations (Jones 1971).

PHYLOGENY OF THE GLUCOSINOLATE-MYROSINASE SYSTEM: CONVERGENCE VS. PARALLELISM

The questions of why and how 12 plant families evolved the capability for glucosinolate and myrosinase synthesis are not amenable to direct experimental attack. Rather, explanations here must rely on the comparative method traditional in systematic research. Nonetheless, experimental methods, as well as the comparative approach, can be used to study glucosinolate diversification at lower taxonomic levels; understanding the processes of diversification at the population and species level allows us to extrapolate to higher taxonomic ranks to explain the origins of the glucosinolate-myrosinase system. If, as argued above, the diversity of glucosinolate composition characterizing various taxa is understandable as the momentary, nonequilibrium outcome of a chemical "arms race" (Feeny 1975) between plants and their herbivores or pathogens, then the origin of the glucosinolate-myrosinase system should prove comprehensible as a biosynthetic shift, in one or a few lineages, from one line of chemical defense to another (cf. Gardner 1977). Such shifts to new classes of chemical defenses may well have occurred <u>within</u> present glucosinolate-producing families. Cardenolides in *Erysimum*, cucurbitacins in *Iberis*, and alkaloids in *Lunaria* may represent novel chemical defenses evolved within the Cruciferae (Feeny 1977; Nielsen 1978). Glucosinolates themselves in the Euphorbiaceae may represent a novel biosynthetic shift in a family more typically defended by alkaloids, cyanogenic glycosides, and triterpenes (see Gibbs 1974: 1343ff). Yet, the biochemical, developmental, and genetic basis for such evolutionary divergence within a family--let alone, among families--is almost wholly unknown. Speculation about the origin or origins of the glucosinolate-myrosinase system must rest on comparative taxonomic study, guided by knowledge of the biosynthetic and developmental pathways of this phytochemical character. Our speculation here is organized around an evaluation of several competing classifications of the families under consideration, three of these schemes being singled out in Table 5 because they are recent and because their authors have used glucosinolates as characters. It is not my intention to introduce yet another classification but to evaluate these current schemes critically and to offer a view of the process of glucosinolate origins that must be reconciled with any phylogenetic classification.

Dahlgren's Classification and Strict Homology

The problem of homology can only be solved with comparative information about the biosynthetic pathways producing the glucosinolates, especially knowledge of the enzymes involved. This is really Cuvier's method applied to chemical characters, instead of anatomical ones. Compared to most classes of secondary plant metabolites, glucosinolate biosynthesis is well known (Kjaer and Larsen 1973, 1976; Tapper and Reay 1973; Underhill et al. 1973), and it appears to be a unitary phenomenon, with synthesis proceeding from precursor amino acids via aldoximes and involving perhaps 10-15 enzymatic steps (Chew and Rodman 1979). This pathway stands on the basis of research with plants from Cruciferae, Resedaceae, and Tropaeolaceae (Underhill et al. 1973). Nonetheless, the critical comparisons that must be made involve the actual enzymes catalyzing the various steps, and little is known about these. In one of the few comparative studies, Matsuo and Underhill (1969, 1971) demonstrated the occurrence of an enzyme with similar activity from *Tropaeolum majus* L. and from three Cruciferae species, which catalyzes the formation of desulfobenzylglucosinolate, a precursor to benzylglucosinolate. No comparative information about structure was reported. Phylogenetically problematic families like Bataceae,

Table 5. Phylogenetic disposition of the glucosinolate-producing families in three recent classifications of flowering plants. Where non-glucosinolate families are included in an order, an "etc." follows the name of the glucosinolate taxa. Names in parentheses pertain to taxa treated under a different family name.

Cronquist (1968)	Thorne (1976)	Dahlgren (1977)
Capparales	Capparales	Capparales
Cruciferae	Brassicaceae	Brassicaceae
Capparaceae	Capparaceae	Capparaceae
Resedaceae	Resedaceae	Resedaceae
Moringaceae	Moringaceae	Moringaceae
Tovariaceae		Tovariaceae
	Cistales	Gyrostemonaceae
Violales	Caricaceae, etc.	Bataceae
Caricaceae, etc.		Salvadoraceae
		Limnanthaceae
	Euphorbiales	Tropaeolaceae
Euphorbiales	Euphorbiaceae, etc.	Bretschneideraceae
Euphorbiaceae, etc.		
	Oleales	Violales
Celastrales	Salvadoraceae, etc.	Caricaceae, etc.
Salvadoraceae, etc.		
	Geraniales	Euphorbiales
Geraniales	Limnanthaceae	Euphorbiaceae, etc.
Limnanthaceae	Tropaeolaceae, etc.	
Tropaeolaceae, etc.		
	Rutales	
Caryophyllales	Gyrostemonaceae, etc.	
(Gyrostemonaceae), etc.		
	Taxa incertae sedis	
Batales	*Batis*, etc.	
Bataceae		

Euphorbiaceae, Gyrostemonaceae, and Salvadoraceae have not been the subject of biosynthetic studies. There is no evidence, therefore, to refute the idea that glucosinolate biosynthesis is an homologous process in these 12 families (see Table 1), but the present data are really too sparse to provide a strong test. Other characters and considerations must be invoked to evaluate the merits of the classifications outlined in Table 5.

The question of homology of the glucosinolate-myrosinase system is not explicitly addressed by either Cronquist (1968) or Thorne (1976) whereas the extreme lumping of Dahlgren (1977) strongly implies that the thioglucosides are homologous throughout the families and orders which he recognizes. Indeed, it was partly in reaction to the possible non-homology and, hence, convergence implicit in Cronquists's system, that Dahlgren offered his unifying proposal (Prof. Dahlgren candidly admitted in a personal communication that his 1977 system was rather an "over-reaction"). Dahlgren's (1977) classification would be compatible with strict homology of the glucosinolates if one were to posit a monophyletic origin of his three orders Capparales, Euphorbiales, and Violales, and further assume that the latter two orders have experienced back or loss mutations restoring the non-glucosinolate condition in all but a few Euphorbiaceae and in all but Caricaceae of his Violales. However, the order Capparales as delineated by Dahlgren (1977) brings together families which are extremely heterogeneous in floral, fruit, and vegetative morphology. Furthermore, it is difficult to reconcile Dahlgren's extreme lumping with the published differences in amino acid sequences of cytochrome c between *Brassica* (Cruciferae) and *Tropaeolum* (Tropaeolaceae). The *Brassica* cultivars rape and cauliflower have identical sequences (Thompson et al. 1971; there may be a minor variant within rape plants, differing by a single amino acid residue: Richardson, et al. 1971), and these differ by nine amino acid positions from that of *Tropaeolum* (garden nasturtium: Brown and Boulter 1974). For comparison, *Brassica* differs from *Triticum* (wheat, Poaceae) in ten amino acid positions (Boulter 1972). In other words, the *Brassica*-*Tropaeolum* difference is nearly as great as the *Brassica* (dicot)-*Triticum* (monocot) difference! It is difficult to evaluate the significance of the cytochrome c data, in part because only ca. 20 species of angiosperms have been studied from a total of ca. 250,000, so perhaps no great weight need be placed on these results. Nonetheless, the amino acid sequence data may eventually come to be viewed as exonerating Cronquists's (1968) suggestion that the Tropaeolaceae (and the related Limnanthaceae) are quite distant from the Capparales.

Cronquist's Classification and Convergence or Distant Parallelism

Cronquist's (1968) treatment is the most extreme in implying convergence or distant parallelism among morphologically quite different plants, but it is not absolutely incompatible with an assumption of homology. That assumption, however, would demand that the presence of glucosinolates represent the retention of a primitive character state (a "symplesiomorphy" in the terminology of Hennig 1966:89) in otherwise evolutionarily derived families. This would appear to be implausible. Glucosinolates are unknown in Cronquist's primitive subclass Magnoliidae, and most of the glucosinolate-producing families are considered by him to be highly specialized and derived, thus requiring--under the assumption of homology--that all intermediate taxa and ancestral forms have gone extinct. Cronquist's scheme (and the rather similar classification of Takhtajan 1969) is more compatible with the view that, overall, the glucosinolates are a class of analogous, not homologous characters and that they represent convergence or parallelism among several phylogenetically distant plant lineages.

The case for homology or, less extremely, for parallelism among a phylogenetically limited subset of advanced dicots appears stronger, however. First, as mentioned above, glucosinolate biosynthesis appears to be similar in all 12 families. The assumption that glucosinolate biosynthetic pathways represent an homologous character set is implicit in the recent discussions by Kjaer and his colleagues (Kjaer et al. 1978; Kjaer and Malver 1979) wherein these workers distinguish evolutionarily primitive states from advanced ones (i.e., biosynthesis from common protein amino acids vs. biosynthesis from chain-extended or otherwise modified amino acid precursors). (These researchers may have wished to distinguish primitive or simple vs. advanced or complex _grades_ of organization or development within sets of non-homologous but analogous characters, but this does not appear to be the intention of the two reports cited here.) Second, certain glucosinolates, like the benzyl, isopropyl, 2-butyl, or isobutyl compounds, derived from the protein amino acids phenylalanine, valine, isoleucine, and leucine, have been identified in at least some species in each of the 12 families (cf. Kjaer et al. 1978; Kjaer and Malver 1979). Third, recent assessments of plant morphology have indicated a closer phylogenetic relationship among Capparales, Caricaceae, and Euphorbiales than is granted in Cronquist's

(1968) system (see Airy Shaw 1966:451; Takhtajan 1969; Thorne 1976; Dahlgren 1975, 1977) and morphological reassessment would also remove the glucosinolate-producing Gyrostemonaceae and Bataceae from Cronquist's isolated subclass Caryophyllidae (see Goldblatt et al. 1976; Carlquist 1978).

Thorne's Classification and Close Parallelism

Thorne's (1976) classification avoids both Dahlgren's (1977) extreme lumping and Cronquist's (1968) distant relationships while allowing for homology or for parallelism among a phylogenetically limited subset of advanced dicots. While one might argue with the particular circumscriptions of his orders, Thorne's alignments have the distinct advantage of allowing for an interpretation of parallel evolution of the glucosinolate-myrosinase system within a limited array of plants. Within such an assemblage one can rationally expect a common biochemical background preadapting for or canalizing the origin of thioglucosides. As argued before (Rodman 1978a), the origin of the glucosinolate-myrosinase system should be sought from groups of plants biosynthetically predisposed in terms of enzymic capabilities (cf. Birch 1973, for a relevant if idiosyncratic discussion). The most likely candidates are plants which synthesized cyanogenic glycosides because the initial steps in the biosynthesis of both glucosinolates and cyanoglycosides, those leading to the formation of aldoximes, are formally identical (Kjaer and Larsen 1973; Underhill et al. 1973). Moreover, cyanogenic glycosides are widespread in the angiosperms and occur, for example, in Cronquist's (1968) primitive subclass Magnoliidae (Seigler 1977; Conn 1979). Thus, cyanogen biosynthesis, particularly the pathway leading to aldoximes, may be a common and ancient feature of many flowering plants. The physiological and ecological background to this hypothesized biosynthetic shift from cyanogens to glucosinolates can most profitably be studied in the Euphorbiaceae since various plants in this family produce cyanogens derived from the amino acids isoleucine and valine (Seigler 1977) while others synthesize glucosinolates from these same precursors. In particular, study of the genus *Drypetes* and its relatives should provide insight into the environmental context of glucosinolate evolution. It is, of course, their unique sulfur metabolism that distinguishes the glucosinolate taxa, and consequently, one should seek their origins from a subset of cyanogenic ancestors which possessed a related sulfur chemistry. In this regard, it may prove more than coincidental that sulfated flavonoids have been identified from several species in the families

Bixaceae, Cistaceae, Frankeniaceae, Guttiferae, and Tamaricaceae (Harborne 1977b; Young this volume), which are placed in or closely allied with the superorder Cistiflorae of Thorne's (1976) treatment, the locus of the glucosinolate-rich Capparales.

In a theoretical discussion, Mayr (1966:609) argued that evolutionary parallelism requires a "common heritage" for the lineages showing a parallel trait. For the glucosinolate-myrosinase system, that common heritage may exist among the series of lineages designated by Thorne's related Cistiflorae, Malviflorae, Geraniiflorae, and Santaliflorae (Thorne 1976). Such a classification brings together plants sharing a complex biosynthetic pathway, including attendant cellular and anatomical features of compartmentation, while preserving a degree of separation among the glucosinolate families which their distinct morphologies would seem to require. In his 1976 treatment, Thorne placed the Gyrostemonaceae (and subsequently the putatively related Bataceae) in his order Rutales, removed from the other glucosinolate families, and thus implied the convergent evolution of a glucosinolate-myrosinase system in this one lineage. The Gyrostemonaceae and Bataceae may indeed stand apart in lacking myrosin cells (Carlquist 1978), but this needs to be tested critically (see discussion above). Unfortunately, analysis of the phylogenetic affinities of these two families has been unsatisfying largely because most of the characters treated have been "absence" characters and little effots has been made to link the two families with others on the basis of shared, derived features (cf. Hennig 1966). Undoubtedly, Thorne's (1976) classification will not prove to be final, but only the next best step in the "unending synthesis" (Constance 1964).

CONCLUSION

Both practical and theoretical interests have nurtured the field of biochemical systematics (to use the title of the seminal work in this area: Alston and Turner 1963). On the practical side, advances in chromatographic techniques have allowed for rapid yet sensitive surveying of the numerous samples that are required in taxonomic projects, with techniques in many cases permitting identification and quantification of the dozens of compounds often encountered in studying a particular class of secondary plant metabolites. A variety of such techniques has been developed for the isolation, separation, identification, and quantification of glucosinolates (see Rodman 1978a for review, including references to electrophoretic studies of myrosinases; Blau et al. 1978 and Nielsen et al. 1979, provide details on the techniques of high pressure liquid chromatography and high voltage paper electrophoresis of glucosinolates, respectively). These techniques have yielded valuable results in taxonomic and biosystematic investigations and in studies on the population biology and ecology of glucosinolate-synthesizing plants.

On the theoretical side, phytochemical characters have attracted interest for two fundamental and interdependent reasons. (1) They provide new kinds of characters that can be used to test existing classifications, which are largely based on comparative anatomy and morphology. Congruence strengthens our respect for the existing classification while non-congruence forces a reappraisal of traditional characters and methods. It was, for example, the evidence of glucosinolate distribution, along with serological findings (Jensen et al. 1964) and morphological reinterpretations (Merxmüller and Leins 1967), that led to the now prevalent view that the old order Rhoeadales was an unnatural juxtaposition of the Papaverales, with affinities to herbaceous Magnoliidae, and the Capparales, with affinities to advanced Dilleniidae (Cronquist 1968). (2) Where biosynthetic pathways and enzymology are known in detail and/or where genetic data are available, phytochemical characters permit rigorous examination of the fundamental problem of convergence, that is, of deciding between homology and analogy. Viewing the glucosinolate taxa overall, the biosynthetic and enzymological data, while among the most extensive for any class of secondary plant metabolites, are sparse but consistent with an interpretation of homology or close parallelism. Assuming the simplest evolutionary scenario, therefore, the classification of Thorne (1976) or of Dahlgren (1977) are to be preferred over that of Cronquist (1968), since the latter treatment is more compatible with multiple origins of analogous glucosinolate-myrosinase systems.

Since the 19th century chemists and botanists have been intrigued with aspects of the glucosinolate-myrosinase system, and ecologists in the 20th century have added a third line of study. The happy commingling of these interests and endeavors has created a fascinating picture of chemical evolution in an ecological setting of plant-heterotroph interactions. Future progress in understanding the evolutionary history of glucosinolate-synthesizing plants will result from a continuing collaboration among ecologists, systematists, and chemists. In the complementarity of their insights and results lies the best hope for discerning the pattern and process of evolution in phytochemical characters.

ACKNOWLEDGEMENTS

I thank Robert Nakamura for his constructive reading of the manuscript, and I gratefully acknowledge financial support from National Science Foundation grants BMS75-03311 and DEB 78-11124 for the chemical studies reported here.

LITERATURE CITED

Ahmed, Z. F., A. M. Rizk, F. M. Hammouda, and M. M. Seif El-Nasr. 1972. Glucosinolates of Egyptian *Capparis* species. Phytochemistry 11:251-256.

Airy Shaw, H. K. 1966. A Dictionary of the Flowering Plants and Ferns. Seventh edition. Cambridge: University Press.

Al-Shehbaz, I. A. 1973. The biosystematics of the genus *Thelypodium* (Cruciferae). Contr. Gray Herb. Harvard Univ. 204:3-148.

Alston, R. E., and B. L. Turner. 1963. Biochemical Systematics. Englewood Cliffs, N. J.: Prentice-Hall.

Andersen, A. S., and R. M. Muir. 1966. Auxin activity of glucobrassicin. Physiol. Plant. 19:1038-1048.

Baker, J. R. 1956. The histochemical recognition of phenols, especially tyrosine. Quart. Jour. Micr. Soc. 97:161-164.

Behnke, H.-D. 1977. Phloem ultrastructure and systematic position of Gyrostemonaceae. Bot. Not. 130:255-260.

Benn, M. 1977. Glucosinolates. Pure Appl. Chem. 49:197-210.

Bergmann, F. 1970. Die glucosinolat-biosynthese im verlauf der ontogenese von *Sinapis alba* L. Zeitschr. Pflanzenphysiol. 62:362-375.

Birch, A. J. 1973. Biosynthetic pathways in chemical phylogeny. Pure Appl. Chem. 33:17-38.

Björkman, R. 1976. "Properties and Function of Plant Myrosinases." In The Biology and Chemistry of the Cruciferae, edited by J. G. Vaughan, A.J. MacLeod, and B. M. G. Jones, pp. 191-205. London: Academic Press.

Blau, P. A., P. Feeny, L. Contardo, and D. S. Robson. 1978. Allylglucosinolate and herbivorous caterpillars: a contrast in toxicity and tolerance. Science 200:1296-1298.

Boulter, D. 1972. The use of comparative amino acid sequence data in evolutionary studies of higher plants. Progr. Phytochem. 3:199-229.

Brown, R. H. and D. Boulter. 1974. The amino acid sequences of cytochrome c from four plant sources. Biochem. Jour. 137:93-100.

Carlquist, S. 1978. Wood anatomy and relationships of Bataceae, Gyrostemonaceae, and Stylobasiaceae. Allertonia 1:297-330.

Cates, R. G. and D. F. Rhoades. 1977. Patterns in the production of antiherbivore chemical defenses in plant communities. Biochem. Syst. Ecol. 5:185-193.

Chan, H. T., R. A. Heu, C. S. Tang, E. N. Okazaki, and S. M. Ishizaki. 1978. Composition of papaya seeds. Jour. Food Sci. 43:255-256.

Chew, F. S. and J. E. Rodman. 1979. "Plant Resources for Chemical Defense." In Herbivores: Their Interaction with Secondary Plant Metabolites, edited by G. A. Rosenthal and D. H. Janzen, pp. 271-307. New York: Academic Press.

Cole, R. A. 1978. Epithiospecifier protein in turnip and changes in products of autolysis during ontogeny. Phytochemistry 17:1563-1565.

------. 1976. Isothiocyanates, nitriles and thiocyanates as products of autolysis of glucosinolates in Cruciferae. Phytochemistry. 15:759-762.

Conn, E. E. 1979. "Cyanide and Cyanogenic Glycosides." In Herbivores: Their Interaction with Secondary Plant Metabolites, edited by G. A. Rosenthal and D. H. Janzen, pp. 387-412. New York: Academic Press.

Constance, L. 1964. Systematic Botany, an unending synthesis. Taxon 13:257-273.

Cronquist, A. 1968. The Evolution and Classification of Flowering Plants. Boston: Houghton Mifflin.

Dahlgren, R. 1975. A system of classification of the angiosperms to be used to demonstrate the distribution of characters. Bot. Not. 128:119-147.

------. 1977. A commentary on a diagrammatic presentation of the angiosperms in relation to the distribution of character states. Plant Syst. Evol. Suppl. 1:253-283.

Dalgaard, L., R. Nawaz, and H. Sørensen. 1977. 3-methylthiopropylamine and (R)-3-methylsulphinylpropylamine in *Iberis amara*. Phytochemistry 16:931-932.

Danielak, R., and B. Borkowski. 1970. New thioglucosides from seeds of dame's violet (*Hesperis matronalis* L.). Diss. Pharmaceut. Pharmacol. 22:143-148.

Davis, P. H., and V. H. Heywood. 1963. Principles of Angiosperm Taxonomy. Princeton, N. J.: Van Nostrand.

Daxenbichler, M. E., C. H. VanEtten, F. S. Brown, and Q. Jones. 1964. Oxazolidinethiones and volatile isothiocyanates in enzyme-treated seed meals from 65 species of Cruciferae. Jour. Agric. Food Chem. 12:127-130.

------, -------, and P. H. Williams. 1979. Glucosinolates and derived products in cruciferous vegetables. Analysis of 14 varieties of Chinese cabbage. Jour. Agric. Food Chem. 27:34-37.

Delaveau, P. 1952. Contribution à l'étude du rôle physiologique du sinigroside de la moutarde noire (*Brassica nigra* Koch., Cruciferes). Compt. Rend. Acad. Paris 234:460-462.

------. 1958. Variations de la teneur en hétérosides à sénevol de l'*Alliaria officinalis* L. au cours de la végétation. Compt. Rend. Acad. Paris 246:1903-1905.

Ehrilich, P. R., and P. H. Raven. 1965. Butterflies and plants: a study in coevolution. Evolution 18:586-608.

Elliott, M. C., and B. B. Stowe. 1971a. Indole compounds related to auxins and goitrogens of woad (*Isatis tinctoria* L.). Plant Physiol. 47:366-372.

----- and -----. 1971b. Distribution and variation of indole glucosinolates in woad (*Isatis tinctoria* L.). Plant Physiol.

48:498-503.

El Migirab, S., Y. Berger and J. Jadot. 1977. Isothiocyanates, thiourées et thiocarbamates isolées de *Pentadiplandra brazzeana*. Phytochemistry. 16:1719-1721.

Estabrook, G. F. 1977. Does common equal primitive? Syst. Bot. 2:36-42.

Ettlinger, M. G. and J. E. Hodgkins. 1956. The mustard oil of papaya seed. Jour. Org. Chem. 21:204.

------ and A. Kjaer. 1968. Sulfur compounds in plants. Rec. Adv. Phytochem. 1:59-144.

------ and A. J. Lundeen. 1956. The structure of sinigrin and sinalbin: an enzymatic rearrangement. Jour. Amer. Chem. Soc. 78:4172-4173.

------ and C. P. Thompson. 1962. Studies of mustard oil glucosides (II). Final Report Contract DA-19-129-QM 1689, Office of Technical Services AD-290 747. Washington, D.C.: U.S. Dept. Commerce.

Fahn, A. 1979. Secretory Tissues in Plants. London: Academic Press.

Feeny, P. 1977. Defensive ecology of the Cruciferae. Ann. Missouri Bot. Gard. 64:221-234.

------. 1976. Plant apparency and chemical defense. Rec. Adv. Phytochem. 10:1-40.

------. 1975. "Biochemical Coevolution Between Plants and Their Insect Herbivores." In Coevolution of Plants and Animals, edited by L. E. Gilbert and P. H. Raven, pp. 3-19. Austin: University of Texas Press.

Flath, R. A. and R. R. Forrey. 1977. Volatile components of papaya (*Carica papaya* L., Solo variety). Jour. Agric. Food Chem. 25:103-109.

Fraenkel, G. 1959. The raison d'etre of secondary plant substances. Science 129:1466-1470.

Gardner, R. O. 1977. Systematic distribution and ecological function of the secondary metabolites of the Rosidae-Asteridae. Biochem. Syst. Ecol. 4:29-35.

Gibbs, R. D. 1974. Chemotaxonomy of Flowering Plants. Vol. 3. Montreal: McGill-Queen's University Press.

Gmelin, R. and A. Kjaer. 1970. Glucosinolates in the Caricaceae. Phytochemistry 9:591-593.

Gmelin, R. and A. I. Virtanen. 1961. Glucobrassicin, der precursor von SCN⁻, 3-indolacetonitril und ascorbigen in *Brassica oleracea* species. Ann. Acad. Sci. Fennica Ser. A II Chem. 107:1-23.

Goldblatt, P., J. W. Nowicke, T. J. Mabry and H.-D. Behnke. 1976. Gyrostemonaceae: status and affinity. Bot. Not. 129:201-206.

Gould, S. J. 1977. Ontogeny and Phylogeny. Cambridge, Mass.: Belknap Press of Harvard University Press.

Greenhalgh, J. R. and N. D. Mitchell. 1976. The involvement of flavour volatiles in the resistance to downy mildew of wild and cultivated forms of *Brassica oleracea*. New Phytol. 77:391-398.

Grob, K. and P. Matile. 1979. Vacuolar location of glucosinolates in horseradish root cells. Plant Sci. Lett. 14:327-335.

Guignard, L. 1893. Recherches sur la localisation des principes actifs chez les Capparidées, Tropéolées, Limnanthées, Resédacées. Jour. Bot Paris 7:345-364, 393-400, 417-426, 444-460.

------. 1890. Recherches sur la localisation des principes actifs des Crucifères. Jour. Bot. Paris 4:385-394, 412-430, 435-455.

Hanover, J. W. 1975. Physiology of tree resistance to insects. Annu. Rev. Entomol. 20:75-95.

Harborne, J. B. 1977a. Introduction to Ecological Biochemistry. London: Academic Press.

------. 1977b. Flavonoid sulphates: a new class of natural

product of ecological significance in plants. Progr. Phytochem. 4:189-208.

von Hayek, A. 1911. Entwurf eines Cruciferen-Systems auf phylogenetischer grundlage. Beih. Bot. Centralbl. 27:127-335.

Hedge, I. C. 1976. "A Systematic and Geographical Survey of the Old World Cruciferae." In The Biology and Chemistry of the Cruciferae, edited by J. G. Vaughan, A. J. MacLeod, and B. M. G. Jones, pp. 1-45. London: Academic Press.

Heinricher, E. 1886. Die eiweissschläuche der Cruciferen und verwandte elemente in der Rhoeadinen-Reihe. Mitteil. Bot. Inst. Graz 1:1-92.

------. 1884. Ueber eiweissstoffe führende idioblasten bei einigen Cruciferen. Ber. dt. Bot. Gesellsch. 2:463-466.

Hennig, W. 1966. Phylogenetic Systematics. Urbana: University of Illinois Press.

Iversen, T.-H. 1970. Cytochemical localization of myrosinase (β-thioglucosidease) in root tips of *Sinapis alba*. Protoplasma 71:451-466.

Janzen, D. H. 1973. Community structure of secondary compounds in plants. Pure Appl. Chem. 34:529-538.

Jensen, U., D. Frohne and O. Moritz. 1964. Serological investigations in the field of Rhoeadales and Ranunculaceae. Serol. Mus. Bull. 32:3-6.

Johansen, D. A. 1940. Plant Microtechnique. New York: McGraw-Hill.

Jones, D. A. 1971. Chemical defense mechanisms and genetic polymorphism. Science 173:945.

Jørgensen, L. B., H.-D. Behnke and T. J. Mabry. 1977. Protein-accumulating cells and dilated cisternae of the endoplasmic reticulum in three glucosinolate-containing genera: *Armoracia, Capparis, Drypetes*. Planta 137:215-224.

Josefsson, E. 1970. Pattern, Content, and Biosynthesis of Glucosinolates in Some Cultivated Cruciferae. Svalöf, Sweden: Swedish Seed Association.

------. 1967. Distribution of thioglucosides in different parts of *Brassica* plants. Phytochemistry 6:1617-1627.

Kjaer, A. 1977. Low molecular weight sulfur-containing compounds in nature: a survey. Pure Appl. Chem. 49:137-152.

Kjaer, A. 1960. Naturally derived isothiocyanates (mustard oils) and their parent glucosides. Progr. Chem. Org. Nat. Prod. 18:122-176.

------. 1963. "The Distribution of Sulphur Compounds." In Chemical Plant Taxonomy, edited by T. Swain, pp. 453-473. London: Academic Press.

------. 1966. "The Distribution of Sulphur Compounds." In Comparative Phytochemistry, edited by T. Swain, pp. 187-194. London: Academic Press.

------. 1974. "The Natural Distribution of Glucosinolates: a Uniform Group of Sulfur-containing Glucosides." In Chemistry in Botanical Classification, edited by G. Bendz and J. Santesson, pp. 229-234. New York: Academic Press.

------. 1976. "Glucosinolates in the Cruciferae." In The Biology and Chemistry of the Cruciferae, edited by J. G. Vaughan, A. J. MacLeod and B. M. G. Jones, pp. 207-219. London: Academic Press.

------ and P. Friis. 1962. Isothiocyanates XLIII. Isothiocyanates from *Putranjiva Roxburghii* Wall. including (S)-2-methylbutyl isothiocyanate, a new mustard oil of natural derivation. Acta Chem. Scand. 16:936-946.

------ and S. E. Hansen. 1958. Isothiocyanates XXXI. The distribution of mustard oil glucosides in some *Arabis* species. A chemotaxonomic approach. Botanisk Tidsskrift 54:374-378.

------ and P. O. Larsen. 1976. Non-protein amino acids, cyanogenic glycosides, and glucosinolates. Biosynthesis 4:179-203.

------ and ------ 1973. Non-protein amino acids, cyanogenic glycosides, and glucosinolates. Biosynthesis 2:71-105.

------, J. Ø. Madsen and Y. Maeda. 1978. Seed volatiles within the family Tropaeolaceae. Phytochemistry 17:1285-1287.

Kjaer, A. and O. Malver. 1979. Glucosinolates in *Tersonia brevipes* (Gyrostemonaceae). Phytochemistry 18:1565.

------, ------, B. El-Menshawi and J. Reish. 1979. Isothiocyanates in myrosinase-treated seed extracts of *Moringa peregrina*. Phytochemistry 18:1485-1487.

Kjaer, A. and A. Schuster. 1972a. Glucosinolates in seeds of *Neslia paniculata*. Phytochemistry 11:3045-3048.

------ and ------. 1972b. Glucosinolates in *Syrenia cana*. Phytochemistry 11:1502-1503.

------, ------, and R. J. Park. 1971. Glucosinolates in *Lepidium* species from Queensland. Phytochemistry 10:455-457.

Kondra, Z. P. and B. R. Stefansson. 1970. Inheritance of the major glucosinolates of rapeseed (*Brassica napus* L.) meal. Canad. Jour. Plant Sci. 50:643-647.

Kutáček, M. and V. I. Kefeli. 1968. "The Present Knowledge of Indole Compounds in Plants of the Brassicaceae Family." In Biochemistry and Physiology of Plant Growth Regulators, edited by F. Wightman and G. Satterfield, pp. 127-152. Ottawa: Runge Press.

Lein, K.-A. 1972. Genetische und physiologische Untersuchungen zur Bildung von Glucosinolaten in Rapssamen: Lokalisierung des Haupt-biosyntheseartes durch Pfropfungen. Zeitschr. Pflanzenphysiol. 67:333-342.

Levin, D. A. 1976. The chemical defenses of plants to pathogens and herbivores. Annu. Rev. Ecol. Syst. 7:121-159.

Lewis, J. A. and G. C. Papavizas. 1971. Effect of sulphur-containing volatile compounds and vapors from cabbage decomposition on *Aphanomyces euteiches*. Phytopathology 61:208-214.

Lüthy, J. and M. H. Benn. 1977. Thiocyanate formation from glucosinolates: a study of the autolysis of allylglucosinolate in *Thlaspi arvense* L. seed flour extracts. Canad. Jour. Biochem. 55:1028-1031.

MacLeod, A. J. and R. Islam. 1975. Volatile flavour components of watercress. Jour. Sci. Food Agric. 26:1545-1550.

Mahadevan, S. 1973. Role of oximes in nitrogen metablism in plants. Annu. Rev. Plant Physiol. 24:69-88.

Matsuo, M. and E. W. Underhill. 1971. Purification and properties of a UDP glucose: thiohydroximate glucosyltransferase from higher plants. Phytochemistry 10:2279-2286.

------ and ------. 1969. A UDP glycose: thiohydroximate glucosyltransferase from Tropaeolum majos L. Biochem. Biophys. Res. Comm. 36:18-23.

Mayr, E. 1966. Animal Species and Evolution. Cambridge, Mass.: The Belknap Press of Harvard University Press.

Merxmüller, H. and P. Leins. 1967. Die Verwandschaftsbeziehungen der Kreuzblütler und Mohngewächse. Bot. Jahrb. Syst. 86:113-129.

Mitchell, J. C. and W. P. Jordan. 1974. Allergic contact dermatitis from the radish, Raphanus sativus. Brit. Jour. Dermatol. 91:183-189.

Nielsen, J. K. 1978. Host plant selection of monophagous and oligophagous flea beetles feeding on crucifers. Entomol. Exp. Appl. 24:562-569.

Nielsen, J. K., L. Dalgaard, L. M. Larsen and H. Sørensen. 1979. Host plant selection of the horse-radish flea beetle Phyllotreta armoraciae (Coleoptera: Chrysomelidae): feeding responses to glucosinolates from several crucifers. Entomol. Exp. Appl. 25:227-239.

Olsen, O. and H. Sørensen. 1979. Isolation of glucosinolates and the identification of o-(α-L-rhamnopyranosyloxy)benzylglucosinolate from Reseda odorata. Phytochemistry 18:1547-1552.

Phelan, J. R. and J. G. Vaughan. 1976. A chemotaxonomic study of Brassica oleracea with particular reference to its relationship to Brassica alboglabra. Biochem. Syst. Ecol. 4:173-178.

Pihakaski, K. and T.-H. Iversen. 1976. Myrosinase in Brassicaceae I. Localization of myrosinase in cell fractions of roots of Sinapsis alba L. Jour. Exper. Bot. 27:242-258.

Procházka, Ž. 1959. Chromatographic proof of the presence of sulforaphene in plantain. Die Naturwiss. 46:426.

Pryor, D. E., J. C. Walker and M. A. Stahmann. 1940. Toxicity of allyl isothiocyanate vapor to certain fungi. Amer. Jour. Bot. 27:30-38.

Richardson, M., J. A. M. Ramshaw and D. Boulter. 1971. The amino acid sequence of rape (*Brassica napus* L.) cytochrome c. Biochim. Biophys. Acta 251:331-333.

Rodman, J. E. 1978a. Glucosinolates: methods of analysis and some chemosystematic problems. Phytochem. Bull. 11:6-31.

------ 1978b. Variation, hybridization, and linkage relations in sea-rockets (*Cakile*, Cruciferae): seed glucosinolate evidence. (abstract) Bot. Soc. Amer. Ser. Misc. Publ. 156: 12.

------ 1976. Differentiation and migration of *Cakile* (Cruciferae): Seed glucosinolate evidence. Syst. Bot. 1:137-148.

------ 1974. Systematics and evolution of the genus *Cakile* (Cruciferae). Contr. Gray Herb. Harvard Univ. 205:3-146.

Rodman, J. E. and F. S. Chew. 1979. Phytochemical correlates of herbivory in a community of native and naturalized Cruciferae. Biochem. Syst. Ecol. 7 (in press).

Schraudolf, H. 1965. Zur Verbreitung von Glucobrassicin und Neoglucobrassicin in höheren Pflanzen. Experientia 21: 520-522.

Schraudolf, H. and H. Weber. 1969. IAN-Bildung ans Glucobrassicin: pH-Abhängigkeit und wachstumsphysiologische Bedeutung. Planta 88:136-143.

Schulz, O. E. 1936. "Cruciferae". In Die natürlichen Pflanzenfamilien, Band 17b, edited by H. Harms, pp. 227-658. Leipzig: Wilhelm Engelmann.

Seigler, D. S. 1977. The naturally occurring cyanogenic glycosides. Progr. Phytochem. 4:83-120.

Sokal, R. R. and F. J. Rohlf. 1969. Biometry. San Francisco:

W. H. Freeman.

Sørensen, H. 1970. o-(α-L-rhamnopyranosyloxy)benzylamine and o-hydroxybenzylamine in *Reseda odorata*. Phytochemistry 9: 865-870.

Takhtajan, A. 1969. Flowering Plants: Origin and Dispersal. Washington: Smithsonian Institution Press.

Tang, C.-S. 1971. Benzyl isothiocyanate of papapa fruit. Phytochemistry 10:117-131.

------. 1973. Localization of benzyl glucosinolate and thioglucosidase in *Carica papaya* fruit. Phytochemistry 12:769-773.

Tang, C.-S., M. M. Syed and R. A. Hamilton. 1973. Benzyl isothiocyanate in the Caricaceae. Phytochmistry 11:2531-2533.

Tang, C.-S. and R. A. Hamilton. 1976. Benzyl isothiocyanate in *Cyclicomorpha* (sic) *solmsii* (Caricaceae). Phytochemistry 15:1767-1768.

Tapper, B. A. and P. F. Reay. 1973. "Cyanogenic Glycosides and Glucosinolates (Mustard Oil Glucosides)". In Chemistry and Biochemistry of Herbage, Vol. 1, edited by G. W. Butler and R. W. Bailey, pp. 447-476. London: Academic Press.

Thompson, E. W., M. Richardson and D. Boulter. 1971. The amino acid sequence of cytochrome c of *Fagopyrum esculentum* Moench (buckwheat) and *Brassica oleracea* L. (cauliflower). Biochem. Jour. 124:783-785.

Thorne, R. F. 1976. A phylogenetic classification of the Angiospermae. Evol. Biol. 9:35-106.

Underhill, E. W., L. R. Wetter and M. D. Chisholm. 1973. Biosynthesis of glucosinolates. Biochem. Soc. Symp. 38:303-326.

VanEtten, C. H., M. E. Daxenbichler, P. H. Williams and W. F. Kwolek. 1976. Glucosinolates and derived products in cruciferous vegetables: Analysis of the edible part from twenty-two varieties of cabbage. Jour. Agric. Food Chem. 24: 452-455.

VanEtten, C. H., M. E. Daxenbichler and I. A. Wolff. 1969. Natural glucosinolates (thioglucosides) in foods and feeds. Jour. Agric. Food Chem. 17:483-491.

Van Etten, C. H. and H. L. Tookey. 1979. "Chemistry and Biological Effects of Glucosinolates". In Herbivores: Their Interaction with Secondary Plant Metabolites, edited by G. A. Rosenthal and D. H. Janzen, pp. 471-500. New York: Academic Press.

------ and ------. 1978. "Glucosinolates in Cruciferous Plants." In Effects of Poisonous Plants on Livestock, edited by R. F. Keeler, K. R. Van Kampen and L. F. James, pp. 507-520. New York: Academic Press.

Vaughan, J. G. and E. I. Gordon. 1973. A taxonomic study of Brassica juncea using the techniques of electrophoresis, gas-liquid chromatography and serology. Ann. Bot. 37:167-184.

------ and J. S. Hemingway. 1959. The utilization of mustards. Econ. Bot. 13:196-204.

Verschaffelt, E. 1911. The cause determining the selection of food in some herbivorous insects. Proc. K. Ned. Akad. Wet. 13:536-542.

Virtanen, A. I. 1965. Studies on organic sulphur compounds and other labile substances in plants. A review. Phytochemistry 4:207-228.

Walker, J. C., S. Morell and H. H. Foster. 1937. Toxicity of mustard oils and related sulfur compounds to certain fungi. Amer. Jour. Bot. 24:536-541.

Werker, E. and J. G. Vaughan. 1976. Ontogeny and distribution of myrosin cells in the shoot of Sinapis alba L.: A light- and electron-microscope study. Israel Jour. Bot. 25:140-151.

------ and ------ 1974. Anatomical and ultrastructural changes in aleurone and myrosin cells of Sinapis alba during germination. Planta 116:243-255.

Whittaker, R. H. and P. P. Feeny. 1971. Allelochemics: chemical interactions between species. Science 171:757-770.

CYANOGENIC COMPOUNDS
AND ANGIOSPERM PHYLOGENY

Stephen G. Saupe

Department of Botany
University of Illinois
Urbana, Illinois 61801

Cyanogens are a class of compounds that liberate hydrogen cyanide (HCN) after chemical or enzymatic hydrolysis. This process is termed cyanogenesis. The intent of this paper is to discuss the chemosystematic potential of cyanogenic compounds for interpreting phylogenetic relationships in the Magnoliophyta.

Unless otherwise indicated, taxa are classified according to the system of Cronquist (1968).

BIOLOGY AND CHEMISTRY OF CYANOGENIC PLANTS

The phenomenon of cyanogenesis in angiosperms has been extensively studied. Several excellent general reviews of the subject have appeared in recent years (Conn 1979b, 1979c, 1980a, 1980b; Seigler 1980a). In addition, many specific aspects of cyanogenic compounds, including their chemistry (Eyjólfsson 1970), coevolution (Jones 1973), function (Jones 1972; Jones, et al. 1978; Ferris 1970), biosynthesis (Conn 1979a), taxonomic significance (Hegnauer 1977a; Paris 1963; Tjon Sie Fat 1979a) and isolation, identification and characterization (Seigler 1975, 1977a) have been reviewed.

Compounds from which HCN is released are derivatives of α-hydroxynitriles (cyanohydrins). Two basic types of cyanogens are recognized depending upon the moiety linked to the hydroxyl group of the carbinol carbon: cyanogenic glycosides and cyanolipids. These compounds are O-β-glycosides or fatty acid esters of cyanohydrins, respectively. The general structure and biogenetic relationship of these compounds is given in Fig. 1. To date, 32 cyanogenic glycosides and lipids have been identified from angiosperms (for structures of these compounds, see Seigler (1977a, 1980a)). Not included in these references are the recently identified cyanogens cardiospermim-5-sulfate (Hübel and Nahrstedt 1979) and linustatin and neolinustatin (Smith, et al. 1980). In addition, two non-cyanogenic nitrile-containing lipids are known (Mikolajczak 1977). These compounds, which are apparently derived from the cyanogenic cyanolipids by an allylic rearrangement do not liberate HCN upon hydrolysis because they are not cyanohydrin derivatives. The corresponding glycoside, sarmentosin, which was recently identified in *Sedum sarmentosum* Bunge. (Fang, et al. 1979), is also non-cyanogenic.

HCN is released from a cyanogen after enzymatic or chemical hydrolysis. Enzymatic hydrolysis is initiated when a β-glycosidase enzyme cleaves the glycosidic linkage yielding a sugar and cyanohydrin which further dissociates to liberate HCN and the parent carbonyl compounent (Conn 1980a). Although this latter reaction can occur non-enzymatically, it is typically catalyzed by a hydroxy-nitrile lyase enzyme. In vivo, this process occurs only after cellular damage, thereby uniting the spatially separated glycosides and hydrolytic enzymes (Conn 1980b). Chemical hydrolysis of cyanogenic compounds is generally accomplished by acid treatment at elevated temperature. The methods utilized for hydrolyzing cyanogenic glycosides has been reviewed elsewhere (Seigler 1975).

Cyanogenic glycosides are biosynthesized from one of several different amino acid precursors by a relatively simple series of biochemical transformations. In summary, the reactions in this pathway involve (in sequence): decarboxylation of the amino acid; oxidation of the α-carbon and its amino group to a nitrile; introduction of oxygen at the β-carbon; and lastly, glycosylation of the α-hydroxynitrile (Conn 1979a). The mechanism of cyanolipid biosynthesis is not known, but these compounds are presumably formed by a pathway that branches off from glycoside biosynthesis at the terminal glycosylation stage.

Five amino acids (leucine, phenylalanine, isoleucine, valine and tyrosine) have been shown to serve as the precursors of cyanogenic compounds (Conn 1979a). Although direct exper-

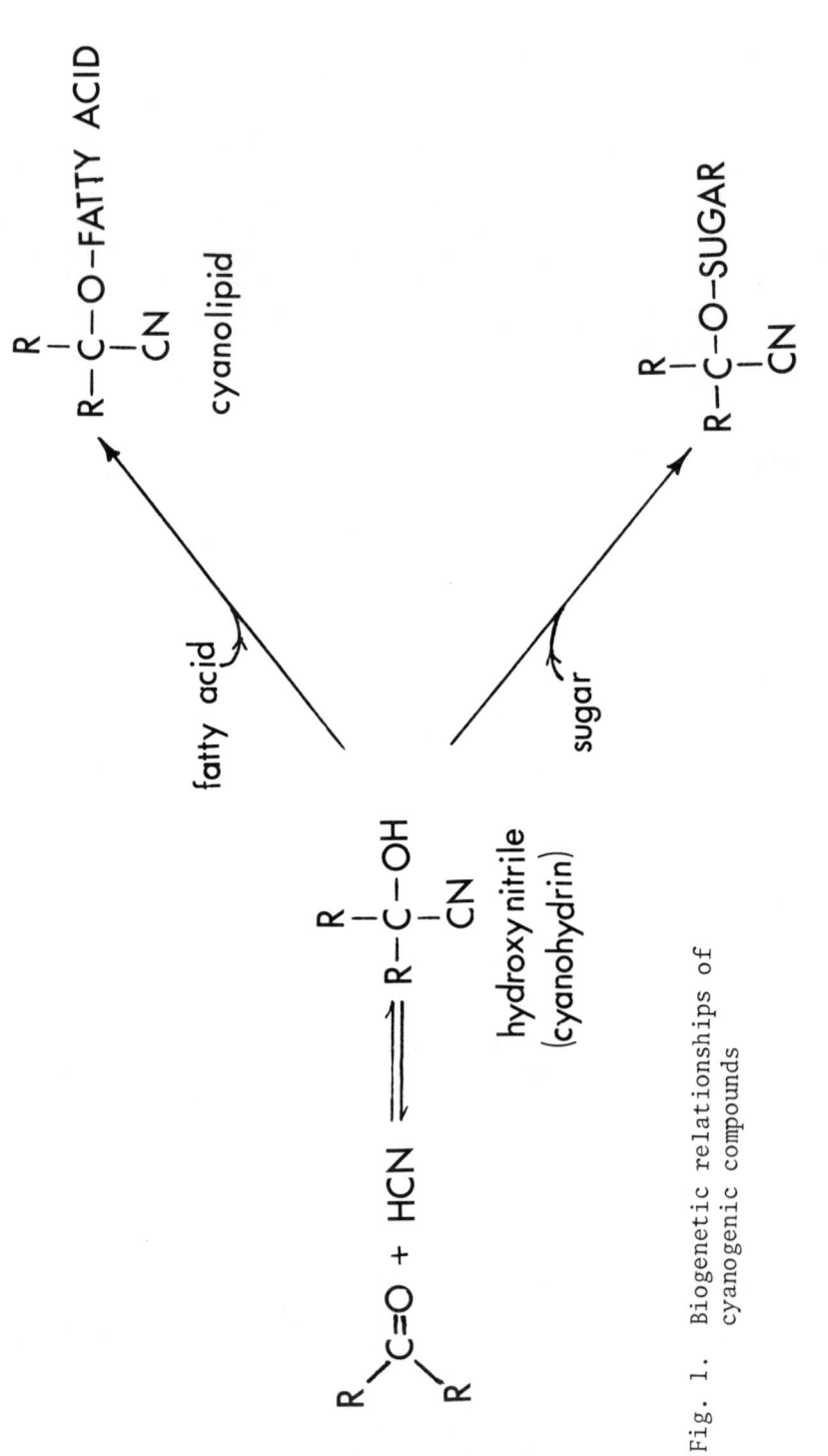

Fig. 1. Biogenetic relationships of cyanogenic compounds

imental evidence for the origin of several cyanogens is lacking, their structural similarity to compounds of known origin strongly suggests their probable precursors. The origin of the cyclopentenoid cyanogens is less certain. These compounds are apparently derived from cyclopentylglycine, a non-proteinaceous amino acid that has been identified in the Flacourtiaceae (Cramer and Spener 1976). Not surprisingly, this family also synthesizes lipids and cyanogenic glycosides that possess the cyclopentene aglycone.

CYANOGENIC COMPOUNDS AS TAXONOMIC CHARACTERS

Cyanogenic compounds are among the many classes of chemical characters (e.g., alkaloids, flavanoids, glycosinolates, proteins, terpenes) that have proven valuable for interpreting phylogenetic relationships in the angiosperms. The taxonomic significance of cyanogens was first recognized in 1830 by Lindley (see Gibbs 1963, 1965) who used HCN production as a trait to distinguish subfamilies of the Rosaceae. A few years later, Endlicher (1836-40; see Gibbs 1965) and Wicke (1851; see Robinson 1930) also applied cyanogenesis as a chemical character in the Rosaceae. One of the earliest applications of cyanogenesis as a character to indicate phyletic lines was by Hallier (1913; see Hegnauer 1959a). Although numerous additions to the catalog of cyanophoric plants were made subsequent to the aforementioned observations, comparatively few chemosystematic investigations during the first half of this century involved cyanogens, and only in the past three decades have cyanogenic compounds been used to any extent in plant classification. This is in large part due to the recent surge of interest in biochemical systematics (Alston and Turner 1963) and to our previously meager knowledge of cyanogens.

A large body of literature has been published in recent years concerning the taxonomic usefulness of cyanogens. Perhaps the most thorough and detailed chemotaxonomic studies involving cyanogenesis have been made by Hegnauer (1962-73) and colleagues (e.g., Fikenscher and Hegnauer 1977a, 1977b; Ruijgrok 1966, 1974; Tantisewie, et al. 1969; Tjon Sie Fat 1979a; van Valen 1979c; and Zandee 1976) at the Experimental Plant Systematics Laboratory in Leiden. No less important are contributions by Gibbs (1974) and others (e.g., Conn 1980a; Eyjólfsson 1968, 1970; Nahrstedt 1973, 1976; and Seigler 1976a, 1976b).

It is generally agreed that cyanogenic compounds are a

Table 1. Hierarchy of taxonomic information derived from cyanogenic compounds.

CHARACTERS	CHARACTER STATES
Qualitative Presence/absence	cyanogenic non-cyanogenic
Structural Configurations	stereochemical variations aglycone variations glycone variations
Biosynthetic Pathways	tyrosine-derived cyanogens leucine-derived cyanogens phenylalanine-derived cyanogens isoleucine/valine-derived cyanogens cyclopentene cyanogens

systematically significant class of chemical characters. Hegnauer (1977a) believes that this class of characters is of considerable importance for the classification of plants and that it can "...reveal tendencies of biochemical evolution". He further suggests that they can furnish evidence to improve existing classification systems. Alston and Turner (1963) and Eyjólfsson (1970) have also recognized the importance of cyanogenesis and cyanogenic compounds in taxonomy. Moreover, Gibbs (1965) argues that HCN can be an important character for deciphering relationships between taxa.

As a class, the cyanogenic compounds provide three major characters applicable to phylogenetic studies. Each character can be further subdivided into two or more character states. The hierarchic organization of chemical evidence from cyanogenic compounds is summarized in Table 1. This classification system provides a logical and comprehensive basis for describing plants and their attributes useful for interpreting relationships between taxa (Radford, et al. 1974). (The phylogenetic and taxonomic utility of each character is discussed in sections IV, V, and VI.)

HCN PRODUCTION AS A CHEMICAL CHARACTER

This character refers to the ability of a plant to liberate HCN without regard for the precursors involved. In essence, it is a qualitative indication of cyanogenic compound formation. Two states of this character exist: a plant is either cyanogenic or non-cyanogenic.

Numerous plants, microorganisms and animals are able to liberate detectable quantities of hydrogen cyanide. This phenomenon occurs mostly in advanced vascular plants (Alston and Turner 1963). More than 2,000 species of angiosperms in approximately 110 families comprising 54 orders possess this biochemical capability (Gibbs 1974; Hegnauer 1977a; Tjon Sie Fat 1979a). Eyjólfsson (1976) and Hegnauer (1958, 1961) estimate that approximately 5% of all angiosperms are cyanogenic. In contrast, only a handful of bacteria (9 species), fungi (30), ferns (50), cyanobacteria (1), algae (3), gymnosperms (6) and insects (30) are cyanogenic (Duffey, et al. 1977; Gewitz, et al. 1974; Harper, et al. 1976; Knowles 1976; Hegnauer 1977a).

HCN is a typical product of metabolism in many families. These include the Araceae, Asteraceae, Calycanthaceae, Euphorbiaceae, Fabaceae, Juncaginaceae, Passifloraceae, Platanaceae, Poaceae, Proteaceae, Rosaceae, Saxifragaceae, and Turneraceae (Conn 1980a; Hegnauer 1959a; Alston and Turner 1963). Although cyanogenesis is known in both monocotyledonous and dicotyledonous angiosperms, this character is most prevalent in dictotyledons. With the exception of the Araceae, Juncaceae, Juncaginaceae, Poaceae and Scheuzeriaceae, cyanogenesis in monocots is rare (Hegnauer 1977a).

Cyanogenesis is a universal character in some taxa. Every species of the Passifloraceae, Malesherbiaceae and Turneraceae tested by Gibbs (1963) was cyanogenic. In addition, all species of several genera such as *Prunus* and *Cotoneaster*, *Aquilegia* and *Isopyrum*, and *Dimorphotheca* liberate HCN (Alston and Turner 1963; Ruijgrok 1966). However, in many taxa, cyanogenesis is not a constant character. In some families, only a single species is cyanogenic. For example, *Tinantia erecta Schlecht.* is the only cyanophoric member of the Commelinaceae (Tjon Sie Fat 1978c). Moreover, closely related species frequently differ in cyanogenic potential. Several species of *Stipa* (Tjon Sie Fat and van Valen 1978), *Lotus* (Jones 1972), and *Campanula* (Tjon Sie Fat 1978a) are cyanogenic while many others are not. Furthermore, individuals in a population may differ in their ability to generate HCN. Polymorphic cyanide production has been recorded for many species including *Trifolium repens* L., *Lotus corniculatus* L., *Manihot esculenta* Crantz., *Prunus amygdalus* Stokes. and *Sorghum bicolor* Pers. (Jones 1972). Recent investigations have suggested that polymorphic HCN production is relatively common (Fikenscher and Hegnauer 1977b).

Cyanogenesis is obviously not a character restricted to closely related taxa. It is a widely distributed process that occurs sporadically in both advanced and primitive angiosperms. Nevertheless, it can provide additional evidence for interpreting phylogenetic problems; particularly in those species which accumulate HCN. Production of cyanide in trace amounts may be the expression of a universal biosynthetic pathway which would be meaningless for taxonomic purposes (Hegnauer 1976, 1977a).

This character can be utilized efficaciously at the ordinal, familial or lower levels (Hegnauer 1977a). However, it is probably of minimal importance at higher taxonomic ranks (Eyjólfsson 1968). It appears to have the greatest value for organizing individual units in a taxon (e.g., subfamilies in a family) (Hegnauer 1959a; Gibbs 1954, 1965). Gibbs (1954, 1965) suggested that the primary use of this character is in groups in which most members are cyanogenic or most are not. Subfamilial taxa that exhibit a high frequency of cyanogenesis (e.g., Fabaceae, Asteraceae) are especially adapted to analysis with this character (Hegnauer 1959a). Cyanogenesis may also be used to study polymorphic species complexes and to typify and identify cultivars (Hegnauer 1959a).

Due to spatial limitations, no specific examples of the chemosystematic application of cyanogenesis will be discussed. The interested reader should consult the comprehensive treatises of Gibbs (1974) and Hegnauer (1962-73).

Cyanogenesis is probably the least valuable of the three characters in this class. The major advantage of this character is that the ease of screening for HCN production has generated a very large data base. Unfortunately, the systematic utility of these data are diminished by several factors:

1. The extensive and sporadic distribution of this phenomenon in angiosperms obscures phylogenetic relationships.
2. Cyanogenesis is exceedingly variable. HCN production is markedly influenced by numerous environmental, developmental and genetic factors (see Jones, et al. 1978 for review).
3. Inaccurately identified specimens and nomenclatural synonymies make the interpretation of literature reports of cyanogenesis difficult (Hegnauer 1977a; Tjon Sie Fat 1979a).
4. The most commonly employed HCN test procedure (picrate) lacks absolute specificity for HCN (Mitchell and Richards 1978). The reported occurrences of cyanogenesis in the Brassicaceae, Droseraceae and several species of Ranunculaceae may be the result of spurious picrate tests (Hegnauer 1977a; Mitchell and Richards 1978; Tjon Sie Fat 1979a).
5. HCN test may fail to detect minute quantities of HCN in weakly cyanogenic plants.

It is apparent that a negative cyanide test cannot discern whether the absence of HCN is due to the lack of the enzymatic and hence, the genetic machinery required for its synthesis, or to a change in the parameters that influence HCN production or its detection. Unless cyanogenesis in a plant is unambiguously determined at the enzymatic level, only positive HCN tests are meaningful for taxonomic purposes (Eyjólfsson 1970). The absence of HCN in a species provides little systematic information (Gibbs 1954).

Even considering the aforementioned limitations, it is apparent that cyanogenesis is a useful character. However, it must be cautiously applied to the interpretation of taxonomic and phylogenetic problems.

STRUCTURAL VARIATION OF CYANOGENIC COMPOUNDS AS A CHEMICAL CHARACTER

The structural configuration of a cyanogenic compound, which is the second major character, may provide several features of potential systematic value. Conn (1980a) recognized three major variations (character states) in the structure of known cyanogens. These are derived from: (1) configuration of the aglycone (e.g., aromatic, aliphatic, alicyclic); (2) nature of the carbohydrate (or lipid) moiety; and (3) stereochemistry of the carbinol carbon.

This character provides a more useful source of taxonomic information than does cyanogenesis. Cyanogen identification permits more definitive conclusions since these data reflect the end-products of an integrated series of enzymatic reactions. Plants with similar cyanogens are predicted to be more closely related than those with dissimilar compounds. However, this is not necessarily true; a cyanogenic compound may occur in two or more phylogenetically unrelated species. For example, triglochinin has been identified in the Araceae and Campanulaceae. One must also exercise caution when applying this character since the origin of several cyanogens is inadequately known. Compounds with different structures may be synthesized via the same pathway in closely related plants; at the same time, metabolites with similar structures may be synthesized by different mechanisms in distantly related species (Seigler 1980b). Hence, just analyzing chemical endproducts may lead to inaccurate conclusions.

The utility of this character is further limited by the small number of plants in which cyanogens have been identified. This is primarily due to the difficulty involved in the isolation and purification of the compounds (Seigler 1977a).

To date, 34 cyanogenic glycosides and cyanolipids have been identified from angiospermous plants of approximately 33 families (Table 2). An individual cyanogen may occur in one or more families. Amygdalin is known only from the Rosaceae. However, its monosaccharide derivative, prunasin, has been

Table 2. Distribution of cyanogenic glycosides in the Magnoliophyta. Unless otherwise indicated, all reports are documented in either Seigler 1977a, 1980a or Tjon Sie Fat 1979, from whom this table is modified and updated. The numbers in parentheses represent the number of species of a genus in which a particular cyanogen occurs. If only a single species is known, its name is given in full.

Biogenetic origin	Compound	Taxon
A. Tyrosine-derived	(S)-Dhurrin	Boraginaceae
		Borago officinalis L.
		Euphorbiaceae
		Bridelia monoica Merrill
		Juncaceae
		Juncus (16, or taxiphyllin)
		Papaveraceae
		Papaver nudicaule L.
		Platanaceae
		Platanus orientalis L. (or taxiphyllin)
		Poaceae
		Sorghum (3)
		Sorghastrum (2) (Gorz, et al. 1979; Haskins, et
		Triticum spelta L. (see Erb, et al. 1979)al 1979)
		Proteaceae
		Macadamia ternifolia F. Muell.
		Stenocarpus sinuatus Endl.
		Ranunculaceae
		Ranunculus (3)
		Trochodendraceae
		Trochodendron aralioides Seib & Zucc.
	p-Glucosoxymandelonitrile	Berberidaceae
		Nandia domestica Thunb.
		Fabaceae
		Goodia lotifolia Salisb.
		Ranunculaceae
		Thalictrum (2)
	Isotriglochinin	Araceae
		Alocasia macrorrhia Schott.
		Juncaginaceae
		Triglochin (2)

(S)-Proteacin   Proteaceae
                    Macadamia ternifolia F. Muell.
                 Ranunculaceae
                    Thalictrum aquilegifolium L.

(R)-Taxiphyllin Calycanthaceae
                    Calycanthus (2)
                    Chimonanthus praecox (L.) Link.
                 Commelinaceae
                    Tinantia erecta Schlect.
                 Euphorbiaceae
                    Phyllanthus gastroemi Muell.
                 Juncaceae
                    Juncus (13, or dhurrin)
                 Juncaginaceae
                    Triglochin maritima L. (Nahrstedt, et al. 1979)
                 Magnoliaceae
                    Liriodendron tulipifera L.
                 Papaveraceae
                    Dicentra spectabilis Lem.
                 Poaceae
                    Bambusa (3)
                    Dendrocalamus (3)

Triglochinin   Araceae
                    Alocasia macrorrhiza Schott.
                    Anthurium hookeri Kunth.
                    Arum maculatum L.
                    Dieffenbachia picta (Lodd.) Schott.
                    Lasia spinosa T. H. W.
                    Pinella tripartita (Blume) Schott.
                 Campanulaceae
                    Campanula (5)
                 Euphorbiaceae
                    Andrachne colchica Fisch et Mey.
                    Bridelia monoica Merrill
                    Poranthera microphylla Brongn.
                    Securinega suffruticosa (Pall.) Rehder
                 Juncaginaceae
                    Lilaea scilloides (Poin.) Haum.
                    Triglochin (2)

Magnoliaceae
   Liriodendron tulipifera L.
Papaveraceae
   Eschscholtzia californica Cham.
   Papaver nudicaule L.
Platanaceae
   Platanus (2)
Poaceae
   Bouteloua hirsuta Lag.
   Chloris (2)
   Cortaderia (6)
   Cynodon dactylon (L.) Pers.
   Elusine (2)
   Glyceria (2)
   Melica (4)
   Molinia caerulea (L.) Moench.
   Siglingia decumbens (L.) Bernh.
   Stipa (3)
   Tridens flavus (L.) Hitch.
Ranunculaceae
   Aquilegia ecalcarata Maxim.
   Leptopyrum fumarioides (L.) Rchb.
   Ranunculus (3)
   Thalictrum polygamum Muhl.
Scheuzeriaceae
   Scheuchzeria palustris L.

B. Leucine-derived   Triglochinin methyl ester

Ranunculaceae
   Thalictrum aquilegifolium L.   (S)-Cardiospermin

Sapindaceae
   Cardiospermum hirsutum Willd.
   Heterodendron oleaefolium Desf.   (S)-Cardiospermin p-hydroxy-benzoate

Rosaceae
   Sorbaria arborea Schneid.

Sapindaceae
   Cardiospermum grandiflorum SW.   Cardiospermin-5-sulfate (Hübel and Nahrstedt 1979)

(S)-Heterodendrin (dihydroacacipetalin)	Fabaceae	
	Acacia sieberiana CD. var. *woodii* (B. Davy) Keay and Brenan	
	Poaceae	
	Holcus mollis L.	
	Sapindaceae	
	Heterodendron olaefolium Desf.	
epi-Heterodendrin	Poaceae	
	Hordeum vulgare L. (Erb, et al. 1979)	
(S)-Proacacipetalin	Fabaceae	
	Acacia (11)	
epi-Proacacipetalin	Fabaceae	
	Acacia globulifera Saff.	

C. Phenylalanine-derived

Amygdalin
    Rosaceae
        Cotoneaster (4)
        Cydonia vulgaris Pers.
        Eriobotrya japonica Lindl.
        Malus (2)
        Photinia serrulata Lindl.
        Prunus (6)
        Sorbus (6)

Holocalin
    Caprifoliaceae
        Sambucus nigra L.
    Fabaceae
        Holocalyx balansae Mich.
    Liliaceae
        Chlorophytum (2)

Lucumin
    Sapotaceae
        Lucuma mammosa Gaertn.

Prunasin
    Asteraceae
        Achillea (2)
        Centaurea scabiosa L.
        Chaptalia nutans (L.) Polak.
    Caprifoliaceae
        Sambucus nigra L.

```
                Fabaceae
                    Acacia (3)
                        Holocalyx balansae Mich.
                Myoporaceae
                        Eremophila maculata (Ker.) F. Muell.
                Myrtaceae
                        Eucalyptus cladocalyx F. Muell.
                Oliniaceae
                        Olinia cymosa Thunb.
                Rosaceae
                        Amelanchier (6) (Majak, et al. 1978)
                        Cotoneaster (32)
                        Cydonia oblonga Brill.
                        Gillenia trifoliata Moench.
                        Prunus (15)
                        Pyracantha coccinea Roem.
                        Sorbus (3)
                        Stranvaesia davidiana Decne.
                Saxifragaceae
                        Jamesia americana Torr. & Gray
                Scrophulariaceae
                        Linaria (2)

   Sambunigrin  Caprifoliaceae
                    Sambucus (2)
                Fabaceae
                    Acacia (3)
                Olacaceae
                    Ximenia americana L.
                Rutaceae
                    Zieria cytosoides Sm. (or prunasin)

   Vicianin     Fabaceae
                    Vicia (5)

   Zierin       Caprifoliaceae
                    Sambucus nigra L.
                Rutaceae
                    Zieria laevigata Sm.
                    Z. cytisoides Sm. (or holocalin)

D. Isoleucine-derived
   Linamarin    Asteraceae
                    Dimorphotheca (6)
```

                    Osteospermum jucundrum Norlindh.
                Euphorbiaceae
                    Cnidoscolus texanus (Muell. Arg.) Small
                    Hevea brasiliensis Muell. Arg.
                    Manihot (3)
                Fabaceae
                    Acacia (2)
                    Lotus (10)
                    Arthinopus (2)
                    Phaseolus lunatus L.
                    Tetragonolobus (2)
                    Trifolium (2)
                Linaceae
                    Linum (4)

Linustatin   Linaceae
                    Linum usitatissimum L. (Smith, et al. 1980)

Lotaustralin Asteraceae
                    Dimorphotheca (2)
                    Osteospermum jacundrum Norlindh.
                Fabaceae
                    Acacia (2)
                    Lotus (10)
                    Orthinopus (2)
                    Phaseolus lunatus L.
                    Tetragonolobus (2)
                    Trifolium (2)
                Euphorbiaceae
                    Manihot carthaginensis Muell. Arg.
                Linaceae
                    Linum (4)

Neolinustatin Linaceae
                    Linum usitatissimum L. (Smith, et al. 1980)

Deidaclin   Passifloraceae
                    Deidamia clematoides (C. H. Wright) Harms
                Turneraceae
                    Turnera ulmifolia L. (Spencer and Seigler, 1980)

Gynocardin   Flacourtiaceae

E. Cyclopentene cyanogens

Gynocardia odorata R. Br.
Pangium edule Reinw.

Tetraphyllin A   Passifloraceae
                 Tetrapathaea tetrandra Cheeseman

Tetraphyllin B   Passifloraceae
                 Adenia volkensii Harms.
                 Barteria fistulosa Mast.
                 Tetrapathaea tetrandra Cheeseman

epi-Tetraphyllin B   Passifloraceae
                       Adenia volkensii Harms.

isolated from the Asteraceae, Caprifoliaceae, Fabaceae, Myoporaceae, Myrtaceae, Oliniaceae, Rosaceae, Saxifragaceae, Scrophulariaceae and several ferns. Likewise, species in a particular family may synthesize one or more cyanogens. Ten cyanogenic compounds are known from the Fabaceae, whereas a single compound (gynocardin) occurs in the Flacourtiaceae.

Of the two structural types of cyanogenic compounds, glycosides and lipids, cyanogenic glycosides are much more common. They occur in numerous, often unrelated plants (Table 2). Hence, the demonstrated occurrence of a cyanogenic glycoside in a species is, in itself, of minimal systematic importance. In contrast, the distribution of cyanolipids is restricted primarily to a single family (Sapindaceae). In this case, it is apparent that these compounds are very useful systematic indicators at the familial level (see VI.c.2).

The aglycones of most cyanogenic glycosides are linked to a monosaccharide (glucose); however, five cyanogens possess a disaccharide glycone. These compounds (and sugar moieties) are amygdalin (gentiobiose), vicianin (vicianose), lucumin (primeverose) and linustatin and neolinustatin (both gentiobiose). The distribution of each of these compounds is restricted to a single family. Vicianin occurs only in the Fabaceae, amygdalin in the Rosaceae, lucumin in the Sapotaceae, and linustatin and neolinustatin in the Linaceae (Table 2). Although much more data are required before the chemosystematic potential of this character state is conclusively known, current evidence suggests that disaccharade cyanogens are characteristic of advanced dicotyledons (Rosidae, Dilleniidae).

Epimeric cyanogens, those which differ in configuration only at the carbinol carbon, often occur in the same family (e.g., taxiphyllin/dhurrin in Poaceae) or genus (e.g., proacacipetalin/epi-proacacipetalin in *Acacia*). However, epimers sometimes occur in species which are not closely related. Heterodendrin is known from the Fabaceae and Sapindaceae while its epimer, epi-heterodendrin, is found only in the Poaceae. Moreover, epimeric cyanogens rarely co-occur. Gondwe, et al. (1978) isolated the epimeric pair tetraphyllin B/epi-tetraphyllin B from *Adenia volkensii* Harms. (Passifloraceae). Sambunigrin/prunasin and holocalin/zierin have been reported from *Sambucus nigra* L. (Caprifoliaceae) (Jensen and Nielsen 1973). Conn (1980a) suggests that the latter report may be an artefact of isolation and purification.

Hübel and Nahrstedt (1979) recently identified cardiospermim-5-sulfate from *Cardiospermum grandiflorum* Sw. (Sapindaceae). The report of this compound, which co-occurs with cardiospermim, is the first of a sulfated cyanogen in plants. Preliminary evidence suggests that the Passifloraceae also synthesize sulfur-containing cyclopentanoid cyanogens (Seigler, D. S. and K. C. Spencer, per. com.). The presently known and restricted distribution of these compounds may be of systematic significance or just an indication of our limited knowledge.

Cyanogenic compounds have been identified in species from each of Cronquist's (1968) recognized subclasses of angiosperms except the Caryophyllidae. A cyanogen of dubious configuration, girgensohnin, has been reported from the Chenopodiaceae (Yurashevskii and Stepanova 1946). Considering the botanical and chemical uniqueness of the "Centrospermae" (Young and Seigler 1980), the unambiguous identification of girgensohnin or other cyanogens from this taxon may prove to be very unusual and may provide additional new evidence for interpreting the "Centrospermae".

CYANOGEN BIOSYNTHESIS AS A CHEMICAL CHARACTER

Of the three aforementioned classes of characters, biosynthesis is the most reliable for interpreting phylogenetic relationships. Phylogenetically, the route of metabolite formation is considerably more meaningful than a specific endproduct (Seigler 1980b). The presence of a biosynthetic pathway in two or more species suggests that they share a common series of enzymes (genes) and therefore are more likely to be related. However, van Valen (1978c) cautions that a pathway of cyanide formation does not necessarily occur in only those species with a narrow phylogenetic relationship. As was true for the preceding character, the taxonomic utility of this character is minimized by the small number of cyanogens known and the even smaller number of plants in which the biosynthesis of the compounds has been demonstrated.

As previously discussed, all cyanogenic glycosides are synthesized from amino acids. Hegnauer (1973), Conn (1980a) and Seigler (1975), who classified the cyanogens according to their amino acid precursors, have established five biogenetic groups (character states). These are: (1) cyanogens derived from tyrosine; (2) cyanogens derived from phenylalanine; (3) cyanogens with a cyclopentene aglycone (probably derived from cyclopentenylglycine); (4) cyanogens derived from leucine; and (5) cyanogens derived from isoleucine/valine. Isoleucine and valine are considered the precursors of a single biogenetic pathway because there is apparently a single set of biosynthetic enzymes which can use either amino acid as a substrate (Conn 1980b).

The origin of cyanogens in each of the major divisions of vascular plants is characteristic for that taxon (Hegnauer 1977a). Cyanogens in the Pteridophyta are derived solely from phenylalanine, whereas gymnosperms only synthesize tyrosine-derived compounds (Hegnauer 1977a). Angiosperms are variable and have been demonstrated to synthesize cyanogens from all five biogenetic groups. It is interesting to note that cyanogenic bacteria and fungi also synthesize HCN from glycine (Knowles 1976).

Within the angiosperms, the distribution of cyanogens based upon their biogenetic origin is characteristic at the familial level. With rare exception, a family synthesizes cyanogens from only a single pathway. The only families presently known with more than one route of cyanogen formation are the Asteraceae, Fabaceae, Euphorbiaceae, Poaceae and Rosaceae. Considering the long recognized chemical and morphological diversity of these large families, this observation is not surprising.

In each of the families with cyanogens of diverse origin, the different pathways are invariably segregated at the infra-familial level. In the Euphorbiaceae, the tyrosine pathway is restricted to the Phyllanthoideae, and the isoleucine/valine pathway occurs exclusively in the Crotonoideae (van Valen 1978e). Likewise, in the Rosaceae, leucine-derived compounds only occur in the Spiraeoideae while phenylalanine-derived compounds are characteristic of the remaining subfamilies (Maloideae, Amygdaloideae) that have been examined (Nahrstedt 1976; van Valen 1978c). Subfamilial segregation of biogenetic pathways also occurs in the Asteraceae (Hegnauer 1977b). Although the distribution of biosynthetic pathways in the Poaceae is not delimited by subfamilial bounds, it is tribally specific (Tjon Sie Fat 1977, 1978b, 1979b; Tjon Sie Fat and van Valen 1978).

Phenylalanine, tyrosine and isoleucine/valine-derived cyanogens are known to occur in the Fabaceae (Table 2). The distribution of these pathways is tribally distinct (van Valen 1979a). A majority of the tribes of the Faboideae (Loteae, Trifolieae, Phaseoleae, Coronilleae) synthesize isoleucine/valine-derived glycosides. A tyrosine-derived cyanogen, which also occurs in this subfamily, is restricted to a separate tribe (Millettieae). The Caesalpinoideae and one tribe of Faboideae (Vicieae) are characterized by the phenylalanine pathway. The genus *Acacia*, the only mimosoid legume from which cyanogens have been identified, is unique because all three character states occur in this taxon. This is the only known subfamilial taxon in which more than one biosynthetic pathway is known to operate (Conn 1980b).

Within the genus *Acacia*, these pathways are distinct; they are separated both geographically and taxonomically. African, Asian and American *Acacias*, which are classified in Bentham's (1875) series Gummiferae, synthesize cyanogens primarily via isoleucine/valine, whereas the cyanogens in Australian species (in the series Phyllodineae, Botrycephalae and Pulchellae) are apparently derived from phenylalanine (Seigler and Conn 1980).

Tyrosine Pathway

Of the five routes of cyanogen formation in angiosperms, the distribution and chemosystematic significance of the tyrosine pathway is the most completely understood. Several tyrosine-derived cyanogens have been reported from angiosperms (Table 2.a.). It is not yet clear if all of these compounds are naturally occurring. Isotriglochinin and triglochinin methyl ester, which have only been isolated together with triglochinin, may be artefacts of chemical work-up (Hegnauer 1977a; Seigler 1980a).

Cyanogens derived from tyrosine have been identified in numerous angiosperms. They are known from approximately 80 species representing 18 families (Table 2.a.). They occur in both monocotyledonous and dicotyledonous angiosperms. In dicots, tyrosine-derived cyanogens occur in the Magnoliidae, Hamamelidae, Rosidae and Asteridae. Moreover, they have been isolated from all of Cronquist's (1968) recognized subclasses of the Liliatae with the exception of the Liliidae. The distribution of tyrosine-derived cyanogens in each of these groups is briefly surveyed in the following discussion.

Monocot cyanogens are derived almost exclusively from tyrosine (Hegnauer 1977a). Taxiphyllin, dhurrin and triglochinin have been isolated from species in the Araceae, Commelinaceae, Juncaceae, Juncaginaceae, Poaceae and Scheuzeriaceae. Although presently available phytochemical data are scant, only two exceptions are known. Epi-heterodendrin (leucine-derived) and holocalin (phenylalanine-derived) were recently isolated from

Hordeum (Poaceae; Erb, et al. 1979) and *Chlorophytum* (Liliaceae; van Valen 1978a), respectively. Although the origin of holocalin in *Chlorophytum* has not been demonstrated, it is presumed to be derived from phenylalanine. However, Hegnauer (1973) and van Valen (1978a) have suggested that holocalin may be synthesized via the tyrosine pathway from m-tyrosine (m-hydroxyphenylalanine). If true, then only one non-tyrosine-derived cyanogen is known in the monocots. However, Seigler (1975) believes that it is more probable that holocalin arises from oxidation of an intermediate derived from phenylalanine than from m-tyrosine.

The Magnoliidae, which is generally recognized as the most primitive group of angiosperms (Cronquist 1968; Thorne 1976), is characterized by the tyrosine pathway of cyanogen formation (Hegnauer 1977a). All known tyrosine-derived cyanogens except isotriglochinin (which is probably an artefact) occur in this subclass. Compounds derived from this pathway occur in the Berberidaceae, Calycanthaceae, Magnoliaceae, Papaveraceae and Ranunculaceae.

Two non tyrosine-derived cyanogens have been reported in this taxon. In 1966, Abrol chromatographically identified linamarin and lotaustralin in the Papaveraceae (*Papaver nudicaule* L.). He also demonstrated that radiolabelled valine and isoleucine were incorporated into these cyanogens. However, van Valen (1978d), who was unable to corroborate Abrol's claims, identified dhurrin and triglochinin from the same species. The occurrence of the isoleucine/valine pathway in the Papaveraceae is unlikely (Hegnauer 1977a; van Valen 1978d).

The Hamamelidae, which is a relatively primitive group of woody dicots with highly reduced flowers, is also characterized by the production of tyrosine-derived cyanogens. Two cyanogenic glycosides have been isolated from species in this subclass. Dhurrin and triglochinin have been isolated from *Trochodendron aralioides* Sieb. & Zucc. (Trochodendraceae; van Valen 1978b) and *Platanus* (Platanaceae; Fikenscher and Ruijgrok 1977), respectively.

Although the tyrosine pathway occurs predominantly in monocots and primitive dicots, it is not wholly restricted to these taxa (van Valen 1979b). It occurs sporadically in the Rosidae (Proteaceae, Euphorbiaceae, Rosaceae) and Asteridae (Campanulaceae, Boraginaceae). It is not known whether these are anomalous occurrences or if this pathway is more widespread than preliminary evidence indicates.

It is generally agreed that the monocots were derived from ancestors resembling primitive dicots (Stebbins 1974). However, there is little agreement as to the ancestral group. Regardless, the distribution of the tyrosine pathway in primitive Magnoliiatae and the Liliatae strongly suggests that these taxa share a common ancestor (Hegnauer 1973).

Leucine Pathway

Leucine is the biogenetic precursor of several cyanogenic glycosides and cyanolipids (Table 2.b.). The distribution and systematic significance of each group of these leucine-derived compounds will be discussed.

1. Cyanogenic Glycosides. Six leucine-derived glycosides have been isolated from angiosperms. The nomenclature and structural configurations of these compounds, which until recently were somewhat controversial, have been reviewed by Seigler (1980a).

Leucine-derived glycosides have a restricted distribution in angiosperms. They are presently known to occur in the Rosaceae, Poaceae, Sapindaceae and *Acacia* (Fabaceae) (Table 2.b.). This pathway is especially prominent in the latter two taxa. With the exception of the Poaceae, families synthesizing cyanogens from leucine are generally considered to be closely related. Cronquist (1968), who places these taxa in the Sapindales (Sapindaceae) and Rosales (Fabaceae and Rosaceae) in the subclass Rosidae, believes that the Sapindales has affinities with the Rosales. Thorne (1976) classifies these taxa in the Rosiflorae (Fabaceae, Rosaceae) and Rutiflorae (Sapindaceae). Recently, Thorne and Dahlgren (see Young and Seigler 1981) support the removal of the Fabaceae from the Rosiflorae to a position in the Rutiflorae. The distribution of the leucine pathway supports the proposed relationship between the Rosaceae, Fabaceae and Sapindaceae. Moreover, it supports Hegnauer's (1977a) suggestion that these taxa belong to the same evolutionary line.

2. Cyanolipids. Mikolajczak (1977) has recently reviewed the biology and chemistry of the cyanolipids. Much of the following discussion is based upon this excellent reference.

The structures of four cyanolipids are known (Fig. 2).

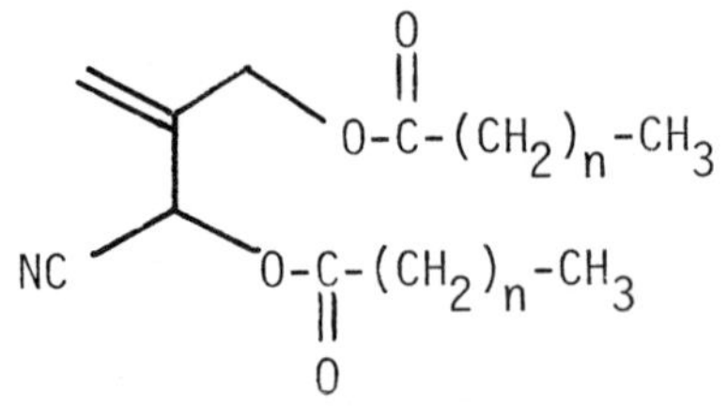

Fig. 2. Structures of cyanolipids occurring in species of Sapindaceae

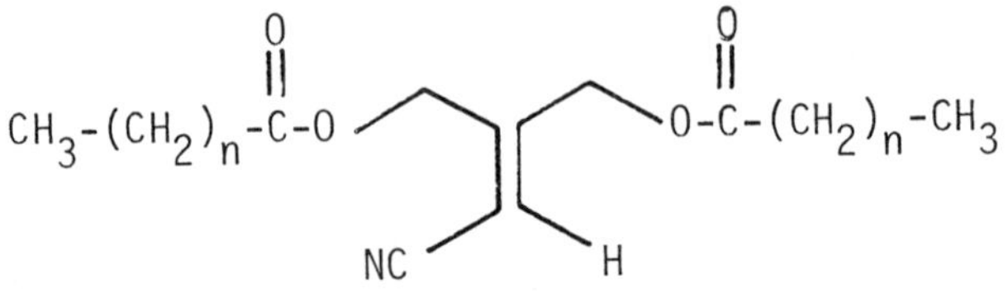

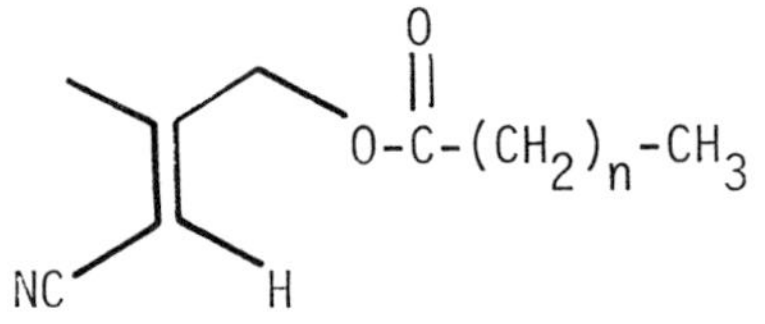

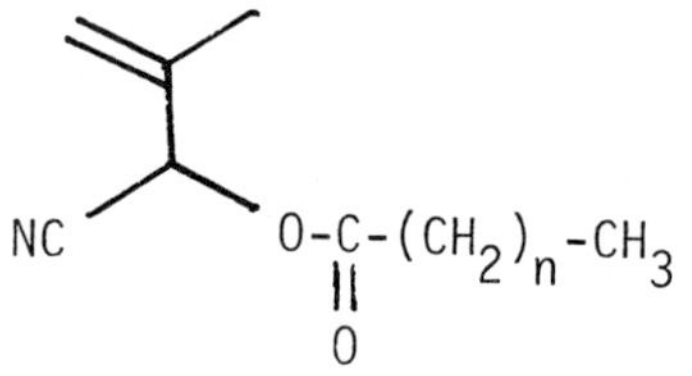

The fatty acid moieties of these compounds vary in chain length and degree of unsaturation. The four cyanolipids have never been demonstrated to co-occur in the same plant. Cyanolipid I is the most common compound while IV is the rarest. The latter compound is known only from *Ungnadia speciosa* Endl. Cyanolipid III always co-occurs with II.

Cyanolipids occur almost exclusively in the seed oils of Sapindaceous plants (Mikolajczak 1977; Seigler and Kawahara 1976). In addition, these compounds have been reported in two genera (*Cordia* (Seigler, et al. 1970); *Heliotropium* (Ahmad et al. 1979)) of the Boraginaceae. Voucher specimens of the *Cordia* collection were subsequently re-examined and determined to be erroneously identified; it was indeed a sapindaceous plant, probably a species of *Allophyllus* (Seigler 1976c). The latter report, which is the only known occurrence of cyanolipids in non-sapindaceaeous plants, warrants confirmation.

The Sapindaceae is unique among angiosperms in its ability to synthesize both structural types of cyanogenic compounds. A species may synthesize cyanolipids (e.g., *Sapindus drummondii* Hook & Arn.) or cyanogenic glycosides (e.g., *Exothea paniculata* Radlk.) or both (e.g., *Cardoispermum hirsutum* Willd.) (Seigler and Kawahara 1976). Other species are completely non-cyanogenic. The co-occurrence of cyanolipids and glycosides in a species suggests that these compounds share similar biosynthetic pathways which apparently branch at the terminal glycosylation stage. Additional supportive evidence for this hypothesis is that the aglycones of cardiospermin and proacacipetalin are identical to those of cyanolipid I and cyanolipid IV, respectively.

Cyanolipid formation is obviously an excellent chemical characteristic of the Sapindaceae. Further, these compounds are absent from closely related families such as the Aceraceae and Hippocastanaceae (Seigler and Kawahara 1976).

Phenylalanine Pathway

Seven cyanogenic glycosides derived from phenylalanine have been isolated from plants. These substances occur in approximately 44 different species representing 13 families (Table 2.c.). The Asteraceae, Fabaceae and Rosaceae are characterized by the production of phenylalanine-derived cyanogens.

The disaccharide cyanogens are the only phenylalanine-derived compounds whose distribution is restricted to a single family. Each of the other compounds derived from this pathway occur in two or more families. Although restricted to a single family, the disaccharide cyanogens may occur in several species; for example, amygdalin is known from 17 Rosaceaeous plants.

The distribution of the phenylalanine pathway is characteristic of advanced Magnoliatae. This pathway occurs almost exclusively in the Rosidae (Fabaceae, Rosaceae, Saxifragaceae, Myrtaceae, Olacaceae, Oliniaceae, Rutaceae) and Asteridae (Myoporaceae, Scrophulariaceae, Asteraceae, Caprifoliaceae), the two most advanced subclasses of angiosperms (Cronquist 1968).

Pteridophytes are also characterized by the phenylalanine pathway of cyanogen formation. Prunasin was isolated from *Cystopteris fragilis* Bernh. and *Pteridium aquilinum (L.)* Kuhn and vicianin occurs in three species of *Davallia* (for references see Seigler 1977a). The disjunct distribution of the phenylalanine pathway in ferns and advanced dicots suggests that this pathway is more ancient than its occurrence in advanced dicots would indicate. Hegnauer (1977a) has recently reviewed the systematic significance of cyanogenesis in ferns.

Isoleucine/valine Pathway

Four cyanogenic glycosides are derived from the isoleucine/valine pathway: linamarin, lotaustralin, linustatin and neolinustatin (Table 2.d.). The latter two compounds, which were recently isolated from flax (*Linum usitatissimum* L.) seed meal, are the diglucoside derivatives of linamarin and lotaustralin, respectively (Smith, Jr., et al. 1979). Valine and isoleucine are the precursors of linamarin and lotaustralin, respectively (Conn 1979a). Although the biosynthesis of linustatin and neolinustatin has not been examined, their structural similarity to linamarin and lotaustralin and their occurrence in a plant in which the isoleucine/valine pathway has been demonstrated, strongly suggest that they also arise from valine and isoleucine.

Linamarin and lotaustralin, with few exceptions, co-occur in a species, although not necessarily in the same ratio (Butler 1965). Linamarin is generally present in the greatest quantity; however, the linamarin:lotaustralin ratio is quite variable between species and individuals (Butler 1965). In addition, linustatin and neolinustatin co-occur in flax.

Butler (1965) and more recently van Valen (1979a) have reviewed the distribution of this character state in higher plants. Glycosides derived from the isoleucine/valine pathway occur in approximately 35 species comprising four families (Table 2.d.). Furthermore, linamarin and lotaustralin were recently isolated from a moth (Davis and Nahrstedt 1979). Isoleucine/valine-derived compounds are particularly characteristic of the Asteraceae, Fabaceae, and Linaceae. This is the only known pathway of cyanogen biosynthesis in the latter family, whereas each of the other families also synthesize cyanogens by alternate pathways.

Cronquist (1968) classifies families that synthesize isoleucine/valine-derived cyanogens in the Rosidae and Asteridae, while Thorne (1976) places them in four superorders (Rosiflorae, Asteriflorae, Malviflorae, Geraniflorae). This pathway appears to have a widespread distribution primarily in advanced, although often unrelated, dicots. These compounds do not appear to be particularly useful as taxonomic characters at the ordinal level or above. However, additional study may prove them to be chemosystematically significant for understanding familial or subfamilial relationships (van Valen 1978a).

Cyclopentene cyanogens

Five cyanogenic glycosides have been isolated that possess a cyclopentene aglycone (Table 2.e.). The absolute stereochemical configuration of these compounds has been determined only for gynocardin (Kim, et al. 1970).

The cyclopentenoid nucleus is not restricted to cyanogenic glycosides; fatty acids with a cyclopentene moiety such as hydnocarpic acid, chaulmoogric acid and gorlic acid, have been isolated from seed oils of Flacourtiaceous plants (Mangold and Spener 1977). These unusual compounds occur without exception in this family.

The cyclopentene cyanogens, which are restricted in distribution to the Flacourtiaceae, Turneraceae and Passifloraceae, are valuable systematic markers (Hegnauer 1977a). Gibbs (1965) and Tantisewie, et al. (1969) have recently reviewed the cyanogenic compound chemotaxonomy of these families.

Based upon chemical and morphological evidence, the Turneraceae is closely related to the Passifloraceae and the Malesherbiaceae is intermediate between the two (see Gibbs 1965). These families are invariably placed in the same order. Thorne (1976) classifies them in the Cistales (Cistiflorae), a small order comprised of 15 families. Cyanogenic families in the Cistales occur in three suborders: Cistineae (Flacourtiaceae); Caricineae (Passifloraceae, Malesherbiaceae, Turneraceae, Caricaceae) and Cucurbitineae (the Cucurbitaceae is dubiously cyanogenic). Every family in the Caricineae for which data are avilable (except Achariaceae) is strongly cyanogenic. Although cyanogens have not been identified in the Malesherbiaceae or Caricaceae, structural analyses will undoubtedly demonstrate that they synthesize glycosides with a cyclopentenoid moiety. Hence, the cyclopentene cyanogens are excellent subordinal characteristics of the Cistales.

Further, the distribution of the cyclopenetene moiety appears to be a relatively good characteristic at the tribal level in the Flacourtiaceae. A species may possess the enzyme systems to synthesize either cyclopentene fatty acids (e.g., *Hydnocarpus anthelmintica* Pierre), glycosides (e.g., *Pangium edule* Reinw., *Gynocardia odorata* R. Br.) or less commonly both types (e.g., *Hydnocarpus ilicifolia* King.). Cyclopentene fatty acids are characteristic of the Oncobeae; cyanogenic glycosides are characteristic of the Pangieae (Hegnauer 1966). There is a minimal degree of biochemical overlap of these pathways between these two tribes. The ability of the Flacourtiaceae to synthesize both lipids and cyanogenic glycosides with similar aglycones is reminiscent of the cyanolipid/cyanoglycoside situation in the Sapindaceae. The biochemistry of cyanogen formation in these families deserves additional study.

ORIGIN AND EVOLUTION OF CYANOGENESIS

Cyanide production is most likely an ancient biochemical phenomena which arose very early in the evolutionary history of life. Eyjólfsson (1976) has recently suggested that cyanogenesis is a chemical shadow of the processes that occurred on this planet countless centuries ago. In addition, he has postulated a theoretical sequence of events for the evolution of cyanogenesis.

Eyjólfsson hypothesizes that HCN, which was an abundant component of the earth's primitive atmosphere, was utilized by certain organisms for their metabolism. This perhaps functioned, at least in part, as a mechanism to detoxify this compound. In response to the depletion of HCN from the atmosphere, cyanide-dependent organisms were compelled to generate their own source of HCN or perish. Some organisms responded by forming cyanogens from amino acids.

The evidence for this hypothesis is convincing and is briefly summarized in the discussion that follows:

(1) HCN is an abundant molecule in the universe. It has been detected in comets, interstellar space, the atmosphere of planets in our solar system, etc. (for references, see Eyjólfsson 1976).

(2) HCN was an important component of the earth's early

atmosphere (Dickerson 1978).

(3) HCN is believed to have played an essential role in the chemistry of pre-biotic and early post-biotic processes, particularly as a catalytic agent in the formation of amino and nucleic acids (Dickerson 1978).

(4) Cyanogenesis is exceedingly widespread in living organisms today. In fact, HCN may be a normal metabolic product of all higher plants (Hegnauer 1977a) and fungi (Locquin 1944). Using sensitive HCN assays, many plants previously recognized as non-cyanogenic by classical techniques have been demonstrated to release HCN (Gewitz, et al. 1974).

(5) Cyanogenic compounds, which have traditionally been considered static waste products, are active metabolites which rapidly turnover in plants (see Conn 1980a, for review). Seigler (1977b) has argued for the role of these compounds in primary metabolism.

(6) Cyanogenic and non-cyanogenic microorganisms (see Knowles 1976, for review) and higher plants (see Conn 1980a; Jones 1972 for reviews) are able to assimilate HCN gas into primary metabolites including alanine, histidine, glutamate, formamide, asparagine and β-cyanoalanine. This ability may serve the dual role of providing precursors for intermediary metabolism and detoxifying HCN.

(7) Cyanide detoxification mechanisms are ubiquitous among cyanogenic and non-cyanogenic organisms. Some organisms metabolize HCN to non-toxic compounds (see #6), an ability that Conn and Butler (1969) suggest arose very early in evolution. Hegnauer (1976) argues that the first stage of secondary metabolite evolution is to develop a physiological tolerance to these generally toxic compounds. Enzymes that detoxify HCN occur in numerous species including photosynthetic bacteria, heterotrophic microorganisms, higher plants and mammals (Knowles 1976; Miller and Conn 1980). Other organisms have evolved HCN-resistant enzymes. Numerous bacteria, fungi and higher plants possess cyanide-insensitive respiratory systems (see Knowles 1976 for review).

Cyanogensis in angiosperms apparently arose very early in evolution. Jones (1972) speculates that "...the genetic systems involved (in cyanogensis) could be very old, perhaps as old as flowering plants". The tyrosine and phenylalanine pathways, which are the only routes of cyanogen formation in gymnosperms and ferns, respectively, probably represent the phylogenetically most ancient pathways of cyanogenesis in angiosperms (Hegnauer 1977a; van Valen 1978b). The antiquity of the tyrosine pathway is further supported by its occurrence as the sole mechanism of cyanogen biosynthesis in primitive dicots and the predominant one in monocots. The absence of cyclopentene and isoleucine/valine-derived cyanogens in these taxa suggest that these pathways are probably of more recent origin, perhaps arising after the divergence of the monocots. Considering the ubiquity of cyanogenesis and that all angiosperms investigated biosynthesize cyanogens from an amino acid by the same pathway, it is probable that cyanogenesis had a monophyletic origin in angiosperms.

LITERATURE CITED

Abrol, Y. P. 1966. Occurrence of linamarin and lotaustralin in Iceland Poppy (*Papaver nudicaule* Linn.). Indian Jour. Chem. 4:251-252.

Ahmad, I., A. A. Ansari and S. M. Osman. 1978. Cyanolipids of Boraginaceae seed oils. Chem. and Ind., 61-2.

Alston, R. E. and B. L. Turner. 1963. Biochemical Systematics. Englewood Cliffs, New Jersey: Prentice-Hall.

Bentham, G. 1875. Revision of the suborder Mimoseae. Trans. Linn. Soc. London 30:335-664.

Butler, G. W. 1965. The distribution of the cyanoglucosides linamarin and lotaustralin in higher plants. Phytochemistry 4:127-131.

------. 1979. "Cyanogenic Glycosides". In Biochemistry of Nutrition. 1A. International Rev. of Biochem. Vol. 27, edited by A. Neuberger and T. H. Jukes, Baltimore, Md.: Univ. Park Press.

Conn, E. E. 1979a. Biosynthesis of cyanogenic glycosides. Naturwissenschaften 66:28-34.

------. 1979b. "Cyanide and Cyanogenic Glycosides". In Herbivores: Their Interactions with Secondary Plant Metabolites, edited by G. A. Rosenthal and D. H. Janzen, New York: Academic Press.

------. 1980a. "Cyanogenic Glycosides". In Secondary Plant Products (Encyclopedia of Plant Physiology; new ser., V. 8) Berlin:Springer Verlag.

------. 1980b. Cyanogenic compounds. Ann. Rev. Plant Physiol. 31:433-451.

Conn, E. E. and G. W. Butler. 1969. "The Biosynthesis of Cyanogenic Glycosides and Other Simple Nitrogen Compounds." In Perspectives in Phytochemistry, edited by J. B. Harborne and T. Swain, London: Academic Press.

Cramer, U. and F. Spener. 1976. Biosynthesis of cyclopentenyl fatty acids. (2-cyclopentenyl)carboxylic acid (aleprolic acid) as a special primer for fatty acid biosynthesis in Flacourtiaceae. Biochem. Biophy. Acta 450:261-65.

Cronquist, A. 1968. The Evolution and Classification of Flowering Plants. Boston: Houghton Mifflin Co.

Davis, R. H. and A. Narstedt. 1979. Linamarin and lotaustralin as the source of cyanide in Zygaena filipendulae L. (Lepidoptera). Biochem. Physiol. 64:395-97.

Dickerson, R. E. 1978. Chemical evolution and the origin of life. Sci. Amer. 239:70-86.

Duffey, S. S., M. S. Blum, H. M. Fales, S. L. Evans, R. W. Roncadori, D. L. Tiemann and Y. Nakagawa. 1977. Benzoyl cyanide and mandelonitrile benzoate in the defensive secretions of millipedes. J. Chem. Ecol. 3:101-113.

Erb, N., H. D. Zinsmeister, G. Lehmann and A. Nahrstedt. 1979. A new cyanogenic glycoside from Hordeum vulgare. Phytochemistry 18:1515-17.

Eyjólfsson, R. 1968. Cyanogenic glycosides in nature, chemistry and distribution. A review. Thesis. The Royal Danish School of Pharmacy, Copenhagen.

------ 1970. Recent Advances in the chemistry of cyanogenic glycosides. Fort. Chem. Org. Natur. 27:74-108.

------ 1976. Cyanogenic compounds and their possible significance concerning the origin of life. J. Brit. Interplanetary Soc. 29:482-48.

Fang, Sheng-din, X. Yan, C. Li, Z. Fan, X. Xu and J. Xu. 1979. Separation and structure of the active principles of Sedum sarmentosum Bunge. K'o Hsueh T'ung Pao 24:431-32 (Chinese).

Chem. Abs. 91:91353q.

Ferris, J. P. 1970. "The Biological Function and Formation of the Cyano Group". In Chemistry of the Cyano Group, edited by E. Rappoport. New York: John Wiley Co.

Fikenscher, L. H. and R. Hegnauer. 1977a. Cyanogenesis bei den cormophyten. 12. Mitteilung. *Chaptalia nutans*, eine stark cyanogene pflanzen Brasiliens. Planta Medica 31:266-269.

------ and ------. 1977b. Die verbreitung der blausäure bei den cormophyten. 11. Mitteilung. Über die cyanogen verbindungen bei einigen Compositae, bei den Oliniaceae und in der Rutaceen-Gattung *Zieria*. Pharm. Weekbld. 112:11-20.

------ and H. W. L. Ruijgrok. 1977. Cyanogenese bei den cormophyten. 13. Mitteilung. Die cyanogenen glucoside von *Platanus* sippen. Planta Medica 31:290-293.

Gewitz, H., G. H. Lorimer, L. P. Solomonson and B. Vennesland. 1974. Presence of HCN in *Chlorella vulgaris* and its possible role in controlling the reduction of nitrate. Nature 249:79-81.

Gibbs, R. D. 1954. Comparative chemistry and phylogeny of flowering plants. Trans. Roy. Soc. Can. 48:1-47.

------. 1954. "History of Chemical Taxonomy." In Chemical Plant Taxonomy, edited by T. Swain. New York: Academic Press.

------. 1965. A classical taxonomist's view of chemistry in taxonomy of higher plants. Lloydia 28:279-299.

------. 1974. Chemotaxonomy of Flowering Plants. Vol. 1-4. Montreal: McGill-Queens University Press.

Gondwe, A., D. S. Seigler and J. E. Dunn. 1978. Two cyanogenic glucosides, tetraphyllin B and epi-tetraphyllin B, from *Adenia volkensii*. Phytochemistry 17:271-274.

Gorz, H. J., F. A. Haskins, R. Dam and K. P. Vogel. 1979. Dhurrin in *Sorghastrum nutans*. Phytochemistry 18:2024.

Harper, N. L., G. A. Cooper-Driver and T. Swain. 1976. A survey for cyanogenesis in ferns and gymnosperms. Phytochemistry 15:1764-1767.

Haskins, F. A., H. J. Gorz and K. P. Vogel. 1979. Cyanogenesis in Indiangrass seedlings. Crop Sci. 19:761-765.

Hegnauer, R. 1958. Over de verspreiding van blauzzuur bij vaatplanten. Pharm. Weekbl. 93:801-819.

------. 1959. "Taxonomic Value of Cyanogenesis in Higher Plants." In Recent Advances in Botany. Vol. I, pp. 82-86. Toronto: University of Toronto Press.

------. 1961. Die verbreitung der blausäure bei den cormophyten. 4. Mitteilung. Untersuchungen über die verbreitung der cyanogenese. Pharm. Weekbl. 96:577-596.

------. 1962-1973. Chemotaxonomie der Pflanzen. Vols. 1-6. Basel: Birkhauser Verlag.

------. 1966. Chemotaxonomie der Pflanzen. Vol. 4. Dicotyledoneae: Daphniphyllaceae-Lythraceae. Basel: Birkhauser Verlag.

------. 1973. Die cyanogen en verbindungen der Liliatae und Magnoliatae-Magnoliidae: Zur systematischen bedeutung des merkmals der cyanogenesis. Biochem. Syst. 1:191-197.

------. 1976. "Accumulation of Secondary Products and its Significance for Biological Systematics." In Secondary Metabolism and Coevolution. Nova Acta Leopoldina. No. 7 Suppl., edited by M. Luckner, K. Mothes and L. Nover, pp. 45-76.

------. 1977a. Cyanogenic compounds as systematic markers in Tracheophyta. Plant Syst. Evol., Suppl. 1:191-209.

------. 1977b. "The Chemistry of the Compositae". In The Biology and Chemistry of the Compositae. Vol. I., edited by V. H. Heywood, J. B. Harborne and B. L. Turner. New York: Academic Press.

Hübel, W. and A. Nahrstedt. 1979. Cardiosperminsulfate-- a sulfur containing cyanogenic glycoside from *Cardiospermum grandiflorum*. Tet. Lett. 45:4395-4396.

Jensen ,S. R. and B. J. Nielsen. 1973. Cyanogenic glucosides in *Sambucus nigra* L. Acta Chem. Scand. 27:1661-1662.

Jones, D. A. 1972. "Cyanogenic Glycosides and Their Function." In Phytochemical Ecology, edited by J. B. Harborne. New York: Academic Press.

------. 1973. "Coevolution and Cyanogenesis." In Taxonomy and Ecology, edited by V. H. Heywood. New York: Academic Press.

------, R. J. Keymer and W. M. Ellis. 1978. "Cyanogenesis in Plants and Animal Feeding." In Biochemical Aspects of Plant and Animal Coevolution, edited by J. B. Harborne. New York: Academic Press.

Kim, H. S., G. A. Jeffrey, D. Panke, R. C. Clapp, R. A. Coburn and L. Long, Jr. 1970. The X-ray crystallographic determination of the structure of gynocardin. Chem. Comm. 381.

Knowles, C. J. 1976. Microorganisms and cyanide. Bact. Rev. 40:652-680.

Locquin, M. 1944. Degagement et localisation de l'acide cyanhydrique chez les basidomyceten et les ascomycetes. Bull. Soc. Linn. Lyon. 13:151-157.

Majak, W., R. J. Bose and D. A. Quinton. 1978. Prunasin, the cyanogenic glycoside in *Amelanchier alnifolia*. Phytochemistry 17:803.

Mangold, H. K. and F. Spener. 1977. "The Cyclopentyl Fatty Acids." In Lipids and Lipid Polymers in Higher Plants, edited by M. Tevini and H. K. Lichtenthaler. Berlin: Springer-Verlag.

Mikolajczak, K. L. 1977. Cyanolipids. Prog. Chem. Fats Other Lipids 15:97-130.

Miller, J. and E. E. Conn. 1980. Metabolism of hydrogen cyanide by higher plants. Plant Physiol. 65:1199-1202.

Mitchell, N. D. and A. J. Richards. 1978. Variation in *Brassica oleracea* L. subsp. *oleracea* (wild cabbage) detected by the picrate test. New Phytol. 81:189-200.

Nahrstedt, A. 1973. Cyanogenesis in *Cottoneaster*-Arten. Phytochemistry 12:1539-1542.

------. 1976. Ein neues cyanogenes glykosid aus der Rosacee *Sorbaria arborea*. Z. Naturforsch. 31:397-400.

------, W. Hösel and A. Walther. 1979. Characterization of cyanogenic glucosides and β-glucosidases in *Triglochin mari-*

tima seedlings. Phytochemistry 18:1137-1141.

Paris, R. 1963. "The Distribution of Plant Glycosides'" In Chemical Plant Taxonomy, edited by T. Swain. New York: Academic Press.

Radford, A. E., W. C. Dickison, J. R. Massey and C. R. Bell. 1974. Vascular Plant Systematics. New York: Harper and Row.

Robinson, M. E. 1930. Cyanogenesis in plants. Biol. Rev. 5:126-141.

Ruijgrok, H. W. L. 1966. "The Distribution of Ranunculin and Cyanogenetic Compounds in the Ranunculaceae." In Comparative Phytochemistry, edited by T. Swain. New York: Academic Press.

------. 1974. Cyanogenese bei *Scheuchzeria palustris*. Phytochemistry 13:161-162.

Seigler, D. S. 1975. Isolation and characterization of naturally occurring cyanogenic compounds. Phytochemistry 14:9-29.

------. 1976a. Plants of the Northeastern United States that produce cyanogenic compounds. Econ. Bot. 30:395-407.

------. 1976b. Plants of Oklahoma and Texas capable of producing cyanogenic compounds. Proc. Okla. Acad. Sci. 56: 95-100.

------. 1976c. Cyanolipids in *Cordia verbenacea*--A correction. Biochem. Syst. Ecol. 4:235-236.

------. 1977a. "The Naturally Occurring Cyanogenic Glycosides." In Progress in Phytochemistry. Vol. 4, edited by L. Reinhold, J. B. Harborne and T. Swain. New York: Pergamon Press.

------. 1977b. Primary roles for secondary compounds. Biochem. Syst. Ecol. 5:195-199.

------. 1980a. Recent developments in the chemistry and biology of cyanogenic glycosides and lipids. Revista Latinoamericana de Quimica (in press).

------. 1980b. "Secondary Metabolites and Plant Systematics."

In The Biochemistry of Plants (Secondary Plant Products). Vol. 7, edited by E. E. Conn. New York: Academic Press. (in press).

------ and E. E. Conn. 1980. Cyanogenesis and taxonomy in the genus *Acacia*. (in preparation).

------ and W. Kawahara. 1976. New reports of cyanolipids from Sapindaceous plants. Biochem. Syst. Ecol. 4:263-265.

------, K. L. Mikolajczak, C. R. Smith, Jr. and I. A. Wolff. 1970. Structure and reactions of a cyanogenetic lipid from *Cordia verbenacea* D.C. seed oil. Chem. Phys. Lipids 4:147-161.

Smith, C. R., Jr., D. Weisleder, R. W. Miller, I. S. Palmer and O. E. Palmer. 1980. Linustatin and neolinustatin: Cyanogenic glycosides of linseed meal that protect animals against selenium toxicity. J. Org. Chem. 45:507-510.

Spencer, K. C. and D. S. Seigler. 1980. Deidaclin in *Turnera ulmifolia* L. Phytochemistry 19:1863-1864.

Stebbins, G. L. 1974. Flowering Plants. Evolution Above the Species Level. Cambridge, Mass.: Belknap Press of Harvard University.

Tantisewie, B., H. W. L. Ruijgrok and R. Hegnauer. 1969. Die verbreitung der blausäure bei den cormophyten. 5. Mitteilung. Über cyanogene vergindungen bei den Parietales und bei einigen Weiteren Sippen. Pharm. Weekbl. 140:1341-1355.

Thorne, R. F. 1976. "A Phylogenetic Classification of the Angiospermae." In Evolutionary Biology. Vol. 9, edited by M. Hecht, W. Steere and B. Wallace. New York: Plenum Press.

Tjon Sie Fat, L. 1977. Contribution to the knowledge of cyanogenesis in Angiosperms. 1. Communication. Cyanogenesis in some grasses (Poaceae (=Graminae)). Proc. Koninkl. Nederl. Akad. Wetensch. Ser. C. 80:227-237.

------. 1978a. Contribution to the knowledge of cyanogenesis in angiosperms. 2. Communication. Cyanogenesis in Campanulaceae. Proc. Konink. Nederl. Akad. Wetensch. Ser. C. 80:126-131.

------. 1978b. Contribution to the knowledge of cyanogenesis in angiosperms. 7. Communication. Cyanogenesis in some grasses. III. Proc. Koninkl. Nederl. Akad. Wetensch. Ser. C. 81:347-354.

------. 1978c. Contribution to the knowledge of cyanogenesis in angiosperms. 8th Communication. Cyanogenesis in Commelinaceae. Lloydia 41:571-573.

------. 1979a. Contribution to the knowledge of cyanogenesis in angiosperms. (Cyanogene verbindingen bij Poaceae, Commelinaceae, Ranunculaceae en Campanulaceae). Diss. R. U. Leiden.

------. 1979d. Contribution to the knowledge of cyanogenesis in angiosperms. 11. Communication. Cyanogenesis in some grasses. IV. The genus *Cortaderia*. Proc. Koninkl. Nederl. Akad. Wetensch. Ser. C. 84:165-170.

------ and F. van Valen. 1978. Contribution to the knowledge of cyanogenesis in angiosperms. 5. Communication. Cyanogenesis in some grasses. II. Proc. Koninkl. Nederl. Akad. Wetenshc. Ser. C. 81:204-210.

van Valen, F. 1978a. Contribution to the knowledge of cyanogenesis in angiosperms. 3. Communication. Cyanogenesis in Liliaceae. Proc. Koninkl. Nederl. Akad. Wetensch. Ser. C. 81:132-140.

------. 1978b. Contribution to the knowledge of cyanogenesis in angiosperms. 4. Communication. Cyanogenesis in *Trochodendron aralioides* Sieb and Zucc. Proc. Koninkl. Nederl. Akad. Wetensch. Ser. C. 81:198-203.

------. 1978c. Contribution to the knowledge of cyanogenesis in angiosperms. 6. Communication. Cyanogenesis in some Magnoliidae. Proc. Koninkl. Nederl. Akad. Wetensch. Ser. C. 81:355-362.

------. 1978d. Contribution to the knowledge of cyanogenesis in angiosperms. 9. Communication. Cyanogenesis in Papaverales. Proc. Koninkl. Nederl. Akad. Wetensch. Ser. C. 81:492-499.

------. 1978e. Contribution to the knowledge of cyanogenesis in angiosperms. 10. Communication. Cyanogenesis in Euphorbiaceae. Planta Med. 34:408-423.

------. 1979a. Contribution to the knowledge of cyanogenesis in angiosperms. XIII. Cyanogenesis in some Papilionaceae. Planta Med. 35:141-145.

------. 1979b. Contribution to the knowledge of cyanogenesis in angiosperms. 12. Communication. Cyanogenesis in Boraginaceae. Proc. Koninkl. Nederl. Akad. Wetensch. Ser. C. 82:171-176.

------. 1979c. Contribution to the Knowledge of Cyanogenesis in Angiosperms. (Cyanogene verbindingen bij Magnoliaceae, Calycanthaceae, Papaveraceae s.l., Trochodendraceae, Papilionaceae, Euphorbiaceae, Boraginaceae en Liliaceae). Diss. R. U. Leiden.

Young, D. A. and D. S. Seigler. 1981. "General Classification and Characteristic of Vascular Seed Plants." In Basic Principles, Vol. 1, Biosolar Resources, edited by A. Mitsui and C. C. Black. West Palm Beach, Florida: CRC Publishing Co. (in press).

Yarashevskii, N. K. and N. L. Stepanova. 1946. Alkaloids of *Girgensohnia oppositiflora*. J. Gen. Chem. (USSR) 16:141-144. (Russ.). Chem. Abs. 40:67548.

Zandee, M. 1976. Beobachtungen uber cyanogenese in der gattung *Juncus*. Proc. Koninkl. Nederl. Akad. Wetensch. Ser. C. 79: 529-543.

TERPENES AND PLANT PHYLOGENY

David S. Seigler

Department of Botany
University of Illinois
Urbana, Illinois 61801

In the past, both chemists and taxonomists have sought to correlate chemical characters (that is, the presence of certain types of compounds) with various taxonomic entities (DeCandolle 1804; Greshoff 1893; McNair 1965; Alton and Turner 1963). Several factors have limited the success of such efforts and it is only in recent years that sufficient data have been available to permit this on a broad scale. In this work, I shall assess the value of chemical data related to the presence and biosynthesis of terpenoids and their relationship to the phylogeny of higher plants. For background information, the reader should consult several previous works: Radford, et al. 1974; Birch 1974; Fairbrothers, et al. 1975; Hegnauer 1976; Seigler 1974, 1977, and in press; Davis and Heywood 1963; and Stebbins 1974.

In the process of evolution, the number and types of proteins and secondary metabolites derived from them have increased. The forces of natural selection have operated on all such products (Stebbins 1974), selecting them for value to and compatibility with parental organisms and the ecological systems in which they occur. Many of these compounds were of a less critical nature than primary metabolites and were less widely distributed. Complications are introduced because one does not observe the primary gene products, but rather pools of compounds they produce, the concentrations of which are partially functions of the relative amounts and activities of enzymes, the availability of certain precursors, and compartmentalization and translocation within the cell (Seigler 1974). Biosynthesis and accumulation of secondary products are distinct processes (Hegnauer 1976). Subsequent mutations may affect steps in a biosynthetic sequence that we observe as an accumulation or disappearance of an altered product. These mutations usually involve the loss, gain, blockage, or alteration of the specificty of an enzyme system. Loss of synthetic ability is presumably more common than gain or alteration, since it merely implies destruction or blocking of a process instead of setting up a new one (Birch 1974). This is partially confirmed by the observation that in several groups of species from the related genera *Parthenium, Hymenoxys,* and *Ambrosia* of the Asteraceae, more highly evolved members have simplified patterns of secondary compounds (Mabry 1974). A one-gene loss may also block an entire pathway.

The determination of homologous origin of similar compounds in different taxonomic groups is one of the fundamental problems inherent in the taxonomic application of secondary compounds (Hegnauer 1967). Two taxa may synthesize or accumulate the same products by different pathways; therefore, the mere presence of a compound is not necessarily an indication of relationship; i.e., similarities in the chemistry of plant taxa (or morphological features) may reflect an evolutionary or phyletic similarity but may also be the result of convergent evolutionary processes (Seigler 1974, 1977).

With a knowledge of biosynthetic pathways of secondary compounds in plants, it should be possible to determine at what point in a sequence divergence has occurred and what subsequent changes have come to pass (Birch 1974). In reality, this is rarely realized because of several factors; several classes of compounds do not appear to have specific structural requirements, whereas in others less variation can be tolerated. For example, most phenolic substances could serve as antioxidants or many lipid compounds for surface coatings as long as the necessary physical properties are met, but attractants for specific pollinators or diterpenes with hormonal activities must be precisely synthesized (Birch 1974). Many plant products arise by simple processes such as removal of activating groups (as phosphate or coenzyme A) or from oxidations, reductions, or methylations of easily modified groups (Birch 1974). In some cases, the relative amounts of products produced may simply reflect the rates of two enzymes operating on a common precursor. Highly probable reactions, such as the introduction of an hydroxyl group ortho or para to an existing one in a phenol, occur frequently in nature. These types of changes are usually of only minor importance in considering the taxonomic significance of secondary compounds.

Other reaction sequences are reversible or are controlled by feedback inhibition such that when a given compound disappears, it disappears without a trace or causes accumulation of a compound far removed in the sequence.

We have limited knowledge as to what pathways may be available in advanced plant groups as we can only see the products of those pathways that the plant utilizes at a particular time. The presence of a given constitutent can only be demonstrated if an appropriate method for its detection is available (Hegnauer 1976).

Several lines of work suggest that many plants are capable of carrying out complex reactions or reaction series but lack precursors or particular enzymes under normal situations. For example, when plants of *Nicotiana* are fed thebaine and certain other precursors of morphine, they are able to perform several biosynthetic steps and produce morphine (Mothes 1966) which is not known to occur naturally in the genus. Interestingly, this conversion cannot be made by some species of *Papaver*, although other species of the genus contain thebaine and morphine.

At the present time, our lack of knowledge of the specific enzymology of the synthesis of secondary metabolites prevents direct comparison of many of the pathways involved in various taxa. Examination and comparisons must frequently be restricted to those systems ascertained to be related by other reasoning, such as a knowledge of the structures of other compounds derived from and part of the biosynthetic pathways in the same and related species of plants.

The forces of natural selection seldom operate on a single organism but on a total biological system. This is undoubtedly one reason convergence in the evolution of both morphological and chemical characters is observed.

A number of chemical reactions are known to occur in virtually all plants. Among these reactions are those which lead to the production of activated compounds such as phosphates and thioesters of coenzyme A. Many enzymes involved in oxidative deamination, other oxidation and reduction processes (such as flavin enzymes, phenol oxidases, peroxidases, dioxygenases, and mixed function oxidases) and enzymes involved in oxidative demethylation appear to be universal. One carbon metabolism involves carboxylases, transcarboxylases, transformylases, hydroxymethyltransferases, methyl transferases (most commonly S-adenosyl methionine) and α-ketoacid decarboxylases. Other important and widespread enzymes are transaminases and amino acid decarboxylases (Luckner 1972; Geissman and Crout 1969). Although these reactions are common to most if not all plants, the enzymes are often found to be specific for a particular substrate or group of substrates. These reactions are not generally useful for taxonomic purposes above the specific or generic level. It is probable that pathways of this type had a common origin and have been inherited by most extant plant groups.

The compounds most useful taxonomically are those which are restricted in distribution. Products which arise by removal of activating groups, oxidation, reduction or methylation of easily modified groups are generally less useful than those which arise via more complex pathways. As a general rule, the more difficult the reactions and the less available the precursors, or the more reaction steps required in a definite sequence to give rise to a compound, the rarer will be its convergent formation (Mothes 1966).

Terpenes are the largest and structurally most diverse of all secondary compounds (Mann 1978). Despite the structural and physical variety of the several thousand compounds of this type, all are united in their biosynthetic origin (Banthorpe and Charlwood 1980). Although it cannot be easily ascertained in certain cases, the structures of all are derived from a

(1)

common precursor, mevalonic acid (1). This six carbon acid gives rise to isopentenyl pyrophosphate and γ,γ-dimethylallyl pyrophosphate which condense to form geranyl, farnesyl, and geranylgeranyl pyrophosphates...the ancestors of monoterpenes, sesquiterpenes, and diterpenes, respectively. Farnesyl and geranylgeranylpyrophosphate dimerize in turn to yield triterpenes and tetraterpenes. Within each series of terpenes, secondary transformations (cyclization, reduction, oxidation, rearrangement) take place (Coates 1976; Mann 1978; Geissman and Crout 1969; and Luckner 1972). Representatives of each of the major groups of terpenes occur in all plants, and thus the basic pathways and precursors of each major group must also occur in each.

In this study, the taxonomic system of Young and Seigler (in press), slightly modified from that of Thorne (1976), has been used as a basis for comparison.

MONOTERPENES

About 400 known monoterpenes occur widely among gymnosperms and angiosperms (Devon and Scott 1972; Banthorpe and Charlwood 1980). These ten carbon compounds usually occur free, but some are found as glycosides or as part of more complex molecules (Charlwood and Banthorpe 1978). They comprise a major part of the volatile constituents or essential oils of plants.

Essential oils are usually isolated from plants by steam distillation, distillation, expression, or extraction. Analysis of a typical oil by capillary gas chromatography indicates that they may consist of as many as 300-400 compounds, although most have only a few major components. In addition to monoterpenes, essential oils frequently contain sesquiterpenes, non-terpenoid esters, aldehydes, ketones, alcohols, hydrocarbons, phenylpropanoids and a variety of other compounds. These oils are of considerable commercial importance as perfume and flavoring ingredients. In nature, they are frequently involved in plant-animal and plant-plant interactions such as pollination, seed and fruit dissemination, allomones, kairomones and allelopathic agents (Harborne 1977; Rosenthal and Janzen 1979).

Essential oils are usually associated with specialized storage structures in plants (Francis 1971). They are frequently found in schizogenous secretory canals, lysigenous cavities, or in special oil cells (Hegnauer 1978).

Problems involving loss or rearrangement of monoterpenes upon isolation by steam distillation are well known (von Rudloff 1969; Charlwood and Banthorpe 1978).

Most monoterpenes are derived by "head to tail" condensation of isopentenyl pyrophosphate (2) and dimethylallyl pyrophosphate (3) and subsequent modification of the geranyl pyrophosphate (4) and neryl pyrophosphate intermediates produced.

(2)

(3)

(4)

The diversity of compounds synthesized (perhaps by unspecific interconversions in a metabolic grid) is influenced by the physiological state of the plant and genetic blocks in individual species or taxa (Banthorpe and Charlwood 1980).

The monoterpenes (and other essential oil components) of many plants have been reported (Hegnauer 1962-1973; Guenther 1948-1959 and biennial reviews in Analytical Chemistry, e.g., Guenther, et al. 1977; Devon and Scott 1972; Tétényi 1970; Karrer, et al. 1977; and numerous articles in Perfumery and Essential Oil Record (which became Flavour Industry in 1969), Journal of Agricultural and Food Chemistry, Journal of the Science of Food and Agriculture, and Phytochemistry.

The analysis of volatile monoterpenes has been of considerable value for the resolution of problems at the generic, specific and infraspecific level (von Rudloff 1975; Adams 1977; and Irving and Adams 1973), but is of limited value for the resolution of phyletic problems. In several, if not most cases, a few genes appear to be involved in the control of structure and amount of monoterpenes produced. The complement of monoterpenes for many plants is often unique, but many of the individual compounds are widely distributed and the pathways leading to them are found in many unrelated taxa of plants.

Although the value of specific compounds and pathways is of limited phylogenetic value, it is clear that certain groups of plants accumulate monoterpenes and possess the morphological structures for storing them. They are often found in gymnosperms (Coniferopsida), the Annoniflorae (order Annonales), Euphorbiaceae, Myrtaceae, Geraniaceae, Rutales (Rutaceae, Meliaceae, Burseraceae, and Anacardiaceae), Juglandaceae, Rosaceae, Pittosporaceae, Araliaceae and Apiaceae, Verbenaceae, Lamiaceae, Asteraceae, Orchidaceae, Rosaceae, Cyperaceae, and Zingiberaceae. They are commonly encountered but occur sporadically or in small amounts in many other families (Hegnauer 1962-1973; Gibbs 1974).

The accumulation of large amounts of monoterpenes in the Annonales but not in the Nelumbonales, Berberidales, and Nympheales supports segregation of that taxon. An overall unity of the superorder is demonstrated by the presence of benzylisoquinoline alkaloids throughout (Seigler 1977).

Dahlgren (1977) placed the Pittosporaceae in his Araliiflorae in agreement with chemical data previously reviewed by Hegnauer (1969). Additionally, Dahlgren segregated the Araliaceae and Apiaceae from the Corniflorae and placed them in his Araliiflorae along with the Pittosporaceae. Both the Araliaceae and Apiaceae lack the iridoid compounds commonly found in the Corniflorae.

Hegnauer (1969, 1978) has pointed out that the Hippuridaceae differs from the Haloragidaceae and Gunneraceae in chemistry and morphology and most closely resembles the Bignoniaceae.

The Verbenaceae and Lamiaceae are quite similar chemically; both accumulate many types of mono-, sesqui-, and di-terpenes and both contain iridoid monoterpenes. They are chemically distinct from the Boraginaceae and Hydrophyllaceae, which are placed in the same superorder by Thorne (1976) and by Young and Seigler (in press).

A group of irregular monoterpenes, among them pyrethrins, occur in several members of the Asteraceae (Heywood, et al, 1977).

IRIDOID MONOTERPENES

In contrast to the common monoterpenes found in essential oils, iridoid monoterpenes occur in all parts of the plants in which they are found and are restricted in distribution. They are not associated with specialized oil storage structures. Iridoids often occur as glycosides and are water soluble (Charlwood and Banthorpe 1978). The chemistry and biosynthesis of this group of compounds has been reviewed (Taylor and Battersby 1969; Banthorpe, et al. 1972; Francis 1971; Cordell 1974; and Charlwood and Banthorpe 1978). Seco-iridoids serve as precursors for indole and monoterpene alkaloids (Cordell 1974).

Although not all steps of the biosynthesis of iridoids have been examined in detail, probable pathways of origin based on a knowledge of precursors and isolated intermediates (i.e., biogenetic pathways) suggest them to be lengthy and complicated. Incorporation studies in this group are often complicated by the presence of side branches to the main pathways (and other factors) which allow incorporation of non-obligatory precursors (Charlwood and Banthorpe 1978).

The distribution of iridoid monoterpenes among higher plant families is restricted to about 40 families, most of which belong to Wettstein's Tubiflorae (Hegnauer and Kooiman 1978; Bate-Smith, et al. 1975) but are found in several other major groups (Table I). Dahlgren (1975, 1977) observed that the presence of iridoid compounds is strongly correlated with the presence of certain morphological features (e.g., unitegmic, generally tenuinucellate ovules) and placed most of the families which have these characteristics in his superorders Corniflorae, Gentianiflorae, and Lamiflorae which he then located closely in his chart of proposed phylogenetic affinities.

Dahlgren (1977) segregated the Araliaceae and Apiaceae and placed them in his Araliiflorae (along with the Pittosporaceae). The biology and chemistry of the Apiaceae have been reviewed (Heywood 1971). These families have acetylenic compounds, petroselinic acid in their seed oils, furocoumarins, and accumulate volatile terpenes, characters not otherwise found in the superorder Corniflorae (Hegnauer 1969). Further,

Table I. Plant Families Known to Contain Iridoid Monoterpenes.

Acanthaceae	Hippuridaceae
Actinidiaceae	Icacinaceae
Adoxaceae	Lamiaceae
Alangiaceae	Lentibulariaceae
Apocynaceae	Loasaceae
Bignoniaceae	Loganiaceae
Buddlejaceae	Martyniaceae
Callitrichaceae	Menyanthaceae
Calyceraceae	Myoporaceae
Caprifoliaceae	Nyssaceae
Cornaceae	Oleaceae
Daphniphyllaceae	Orobanchaceae
Dipsacaceae	Pedaliaceae
Ericaceae	Plantaginaceae
Escalloniaceae	Proteaceae
Eucommiaceae	Rubiaceae
Fouquieraceae	Sarraceniaceae
Garryaceae	Saxifragaceae (*Deutzia*, *Hydrangea*)
Gentianaceae	
Globulariaceae	Scrophulariaceae
Hamamelidaceae (*Liquidambar*)	Valerianaceae

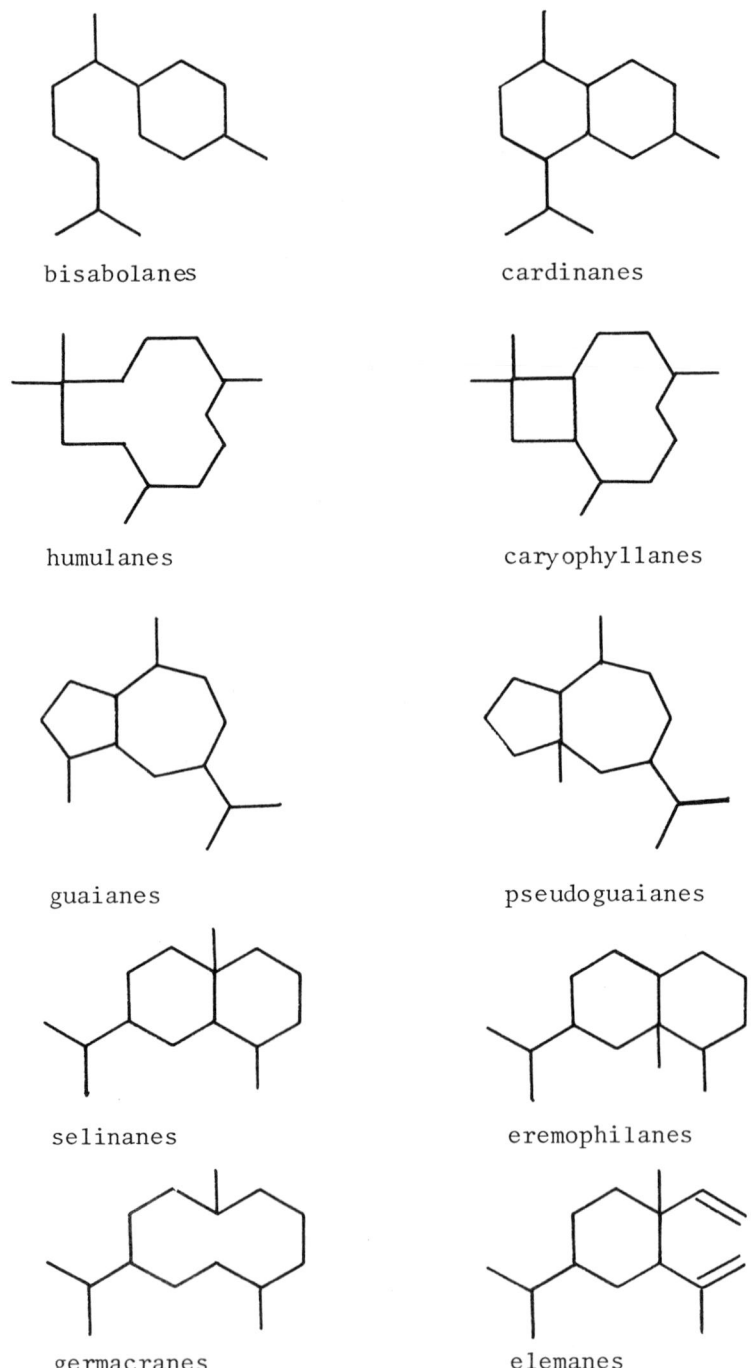

Figure 1: Common Skeletal Types of Sesquiterpenes.

they lack iridoid monoterpenes. The presence in the Apiaceae of sesquiterpene lactones, benylisoquinoline alkaloids (Gupta, et al. 1976), furocoumarins and volatile terpenes suggests alliances with the Asteriflorae and Rutiflorae (Meeuse 1970; Dahlgren 1977; Kubitzski 1969). Cronquist also separated these two orders (1968).

Dahlgren, et al. (1976) placed the Fouquieriaceae in his Corniflorae. The Fouquieriaceae is the only family in the superorder Solaniflorae to contain iridoid terpenes.

In the order Bignoniales, iridoid monoterpenes have been reported from all families.

Chemically, the Gentianales and Bignoniales are similar in that both contain iridoid compounds but distinct in that alkaloids are only common in the former order.

The order Gentianales is characterized by the presence of iridoid monoterpene glycosides and indole alkaloids which are present in all families except the Asclepiadaceae and possibly Buddlejaceae (Hegnauer and Kooiman 1978). Alkaloids are found in the Rubiaceae (emetine, quinine, indole types), the Gentianaceae (monoterpene type), the Loganiaceae, and Apocynaceae (indole and steroidal types). The Asclepiadaceae contain tylophorine alkaloids which are unique to that family (Seigler 1977).

SESQUITERPENES

Sesquiterpenes make up the largest group of terpenoid compounds. The approximately 1,000 known compounds can mostly be grouped into about 30 skeletal types, but at least 70 less common skeletal types are known (Devon and Scott 1972). The distribution of sesquiterpenes is about the same as that of monoterpenes. Sesquiterpene hydrocarbons are common essential oil components. Abscisic acid, a plant growth compound, presumably occurs in all higher plants.

Sesquiterpenes are derived from a fifteen carbon intermediate, farnesyl pyrophosphate. Biogenetic paths to most structural types have been proposed, but few have been studied in detail. Characteristic of all is an initial intramolecular attack to produce a monocyclic system which is governed by either electronic or steric influences. Subsequent 1,2- and

1,3-hydride shifts, oxidations, cyclizations and reductions lead to the observed diversity of structures (Herout 1973; Rücker 1973; Cordell 1976). Not surprisingly, this diversity is greater than for monoterpenes, but it is also greater than for diterpenes and triterpenes. This situation presumably reflects the much greater conformational flexibility of the ten or eleven membered rings as compared to that of six membered rings in the last two classes (Charlwood and Banthorpe 1978). Both 2-trans-6-trans-farnesyl (5) and 2-cis-6-trans-farnesyl (6) pyrophosphate have been thought to serve as intermediates for various types of cyclic systems. Although the conversion of

(5)

(6)

the 2-trans to 2-cis farnesyl pyrophosphate has been demonstrated in some systems, there is also evidence for the direct conversion of the 2-trans compound to cyclic intermediates without the intermediacy of the 2-cis compound, possibly because the conformational flexibility of the ten carbon ring permits reaction (Banthorpe and Charlwood 1980).

Despite the fact that both sesquiterpenes and monoterpenes are isolated together in essential oils, there is evidence that they are not synthesized at the same site (Banthorpe and Charlwood 1980).

Although the number of acyclic sesquiterpenes is small, cis- and trans-farnesol, nerolidol and corresponding unsaturated compounds are relatively common in essential oils.

The monocyclic humulanes (eleven membered ring) and the bicylcic caryophyllanes (cyclobutane-cyclononane) probably do not arise from a common precursor as previously thought. β-Caryophyllene probably arises from a 2-cis whereas α-humulene arises from a 2-trans cation. Both are widely distributed in nature (Rücker 1973). The same precursor which leads to caryophyllene also leads to longifolene, longiborneol and longicyclene, all of which are found in pine resins (Rücker 1973).

The 2-cis-6-trans precursor gives rise to monocyclic types with one six membered ring (bisabolenes) and systems with a 1,7-dimethyl-4-isopropyl skeleton (Fig. 1). Among these are the cadinanes, cubebanes, copabornanes, copaanes, and ylangenes (Rücker 1973).

A complex type of sesquiterpene known as picrotoxins is derived from copaborneol. These compounds are highly toxic and contain a lactone ring. They are found in the Menispermaceae (*Menispermum, Cocculus,* and *Anamirta*), Coriariaceae, Euphorbiaceae (*Hyenanche*) and Orchidaceae (*Dendrobium*) (Coscia 1969).

Germacranes, usually derived from a 2-trans-6-trans intermediate, possess a 4,10-dimethyl-7-isopropyl cyclodecane system and often have double bonds in positions 1(10) and 4 (Fig. 1). The reactive cyclodecadiene system can be transferred readily into other mono, bi, and tricyclic sesquiterpenes and thus the germacranes are key intermediates in the synthesis of other sesquiterpenes. Other major types are the selinanes (eudesmanes), guaianes, pseudoguaianes, elemanes, and eremophilanes (Rücker 1973).

Guaianes, eremophilanes, and pseudoguaianes are especially common in the Asteraceae, the last type occurring only as lactones and in that family (Heywood, et al. 1977).

The same phylogenetic conclusions for monoterpenes are true for sesquiterpenes. Despite their diversity, these compounds have proven of limited value in establishing the phylogeny of higher plants.

SESQUITERPENE LACTONES

Lactones (cyclic esters) of many different types of terpenes are known to occur, but they are most numerous and best studied in the sesquiterpene series (Yoshioka, et al. 1973; Fischer, et al. 1979). These compounds possess potent biological activity and are responsible for the bitter taste and toxic properties of many of the plants in which they occur. They are restricted in distribution, occurring primarily in the Annoniflorae (orders Annonales and Berberidales), the Apiaceae, the Asteraceae and liverworts (Hepaticae). They are most common in the Asteraceae.

Although many details of the biosynthesis of sesquiterpene lactones are not established, it appears as though guaianolides, eudesmanolides, and pseudoguaianolides in the Asteraceae are derived from germacranolide precursors (Fischer 1978; Fischer, et al. 1979).

Although sesquiterpene lactones have proven to be of limited value for study of phylogeny at the higher categorical levels, they have proven useful at the tribal, generic and specific level in the Asteraceae. The biology and chemistry of this family have been reviewed (Heywood, et al. 1977).

DITERPENES

At least 650 diterpenes are found in a large number of plant families (Devon and Scott 1972; Banthorpe and Charlwood 1980). Most of these compounds fit into about 20 major groups. Because they are chemically complex, difficult to isolate, purify and characterize, they have probably not been studied as thoroughly as mono- and sesquiterpenes. The acyclic phytol side chain of chlorophyll and gibberellins occur in all plants.

Diterpenes are usually components of plant resins and are sometimes encountered as by-products from the isolation of essential oils (e.g., from turpentine production).

Several monocyclic diterpenes occur in nature. Cembrene (7), found in the resinous material from several gymnosperms, is the precursor of taxanes which are only found in the Taxaceae. Taxane alkaloids (derived from these diterpenes) are re-

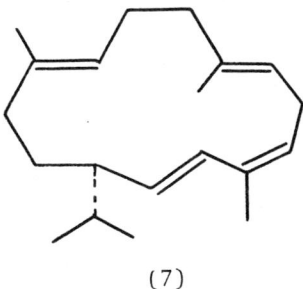

(7)

sponsible for the extremely poisonous properties of the genus *Taxus*.

Other series of monocyclic diterpenes occur in angiosperms. Most important among these are duvatrienediol from tobacco, casbene (from *Ricinus*), and casbene derivatives (e.g., 8) from the Euphoribaceae and Thymeliaceae. The toxic effects and distribution of these compounds have been reviewed (Kinghorn 1979; Hecker 1971). The complexity of these structures and their limited distribution suggest a close relationship between the Euphorbiaceae and the Thymeliaceae.

Two distinct modes of cyclization lead to bicyclic diterpenes with a 5-α-10-β configuration or a 5-β-10-α configuration (Mann 1978; Hanson 1972; Hanson 1971). Some plants are known which have diterpenes with both configurations (e.g., *Agathis australis* (Araucariaceae)) and in other examples different species of the same genus possess opposite configurations (e.g., *Podocarpus macrophyllus* and *P. ferrugineus*, (Podocarpaceae) (Oehlschlager and Ourisson 1967). Diterpenes with the 5-α-10-β configuration are common in gymnosperms but are also known from the Fabaceae, Lamiaceae, Verbenaceae, Rubiaceae, and Asteraceae (de Mayo 1959; MacMillan 1971).

Subsequent modification of the precursor to both 5-α-10-β and 5-β-10-α bicyclic diterpenes may lead to diterpenes which differ in configuration at position 13 (e.g., sclareol and 13-episclareol from *Salvia sclarea*. Furan formation between C-15 and C-16 is a common feature of several groups of plants, notably the Lamiaceae, Verbenaceae, Euphorbiaceae, and Asteraceae.

Tricyclic diterpenes fall into four main classes: those with the pimarane, abietane, rosane, or cassaine skeleton. Epimers at C-13 exist among many groups of tricyclic diterpenes (e.g., pimaric acid and sandaracopimaric acid).

Many pimarane and abietane derivatives (mostly with 5-α-10-β configuration) (e.g., abietic acid (9)) are found in gymnosperms but are also known from the Asteraceae, Lamiaceae, Euphorbiaceae, and also the genus *Jungermannia* (Hepaticae). Compounds of the rosane type occur in the wood of *Erythroxylum* species (Erythroxylaceae) and those of the cassaine type in the Fabaceae (subfamily Caesalpinoideae). *Erythrophleum* alkaloids possess the last type of skeleton (Nakanishi, et al. 1974).

Most tetracyclic diterpenes possess the 10-α-5-β configuration such as occurs in ent-kaurene (10), an intermediate in the

(8)

(9) (10)

biosynthesis of gibberellins. These compounds appear to be widely disbributed but are especially common in the Euphorbiaceae (e.g., in *Beyeria*), the Asteraceae (e.g., in *Stevia*) and the gymnosperms.

A modified type of ent-kaurene system is found in the diterpene alkaloids of the Garryaceae and the Ranunculaceae.

The gibberellins (ubiquitous in plants) (MacMillan 1971), grayanotoxins (Ericaceae) and enmeins (*Isodon*, Lamiaceae) arise by migration of bonds in ent-kaurene precursors (Newman 1978).

Ginkgolides are only known to occur in *Gingko biloba* (Nakanishi, et al. 1974).

Diterpenes appear to be useful for phylogenetic studies of several groups of plants. Among these are cocarcinogenic diterpenes of the Euphorbiaceae and Thymeliaceae, grayanotoxins of the Ericaceae, cembrene and its relatives in gymnosperms, and taxanes of the Taxaceae.

The bicyclic diterpenes of the Lamiaceae, Verbenaceae, Euphorbiaceae and Asteraceae appear to have exceptional promise at the familial and generic level but more study is needed.

TRITERPENES

Triterpenes form a large group of lipid substances which are found in all plants. The structures of the approximately 750 known tripterpenes may be subdivided into about 30 major types. Steroids are derivatives of triterpenes; this group of compounds is also ubiquitous and may be divided into about 20 structural types (Devon and Scott 1972). Although long studied, correct structures for many were not forthcoming until about 1940 (Connolly and Overton 1972). In contrast to the previously discussed mono-, sesqui- and diterpenes, which are biosynthesized by linear condensation of five carbon units, triterpenes arise via dimerization of two farnesyl pyrophosphate units to produce an intermediate compound, squalene (11). Squalene may then be cyclized by several mechanisms in different conformations on an enzyme surface to produce major lines of triterpenes (Fig. 2) which undergo subsequent modification (Banthorne and Charlwood 1980).

Several pentacyclic triterpene hydrocarbons (e.g., fernane,

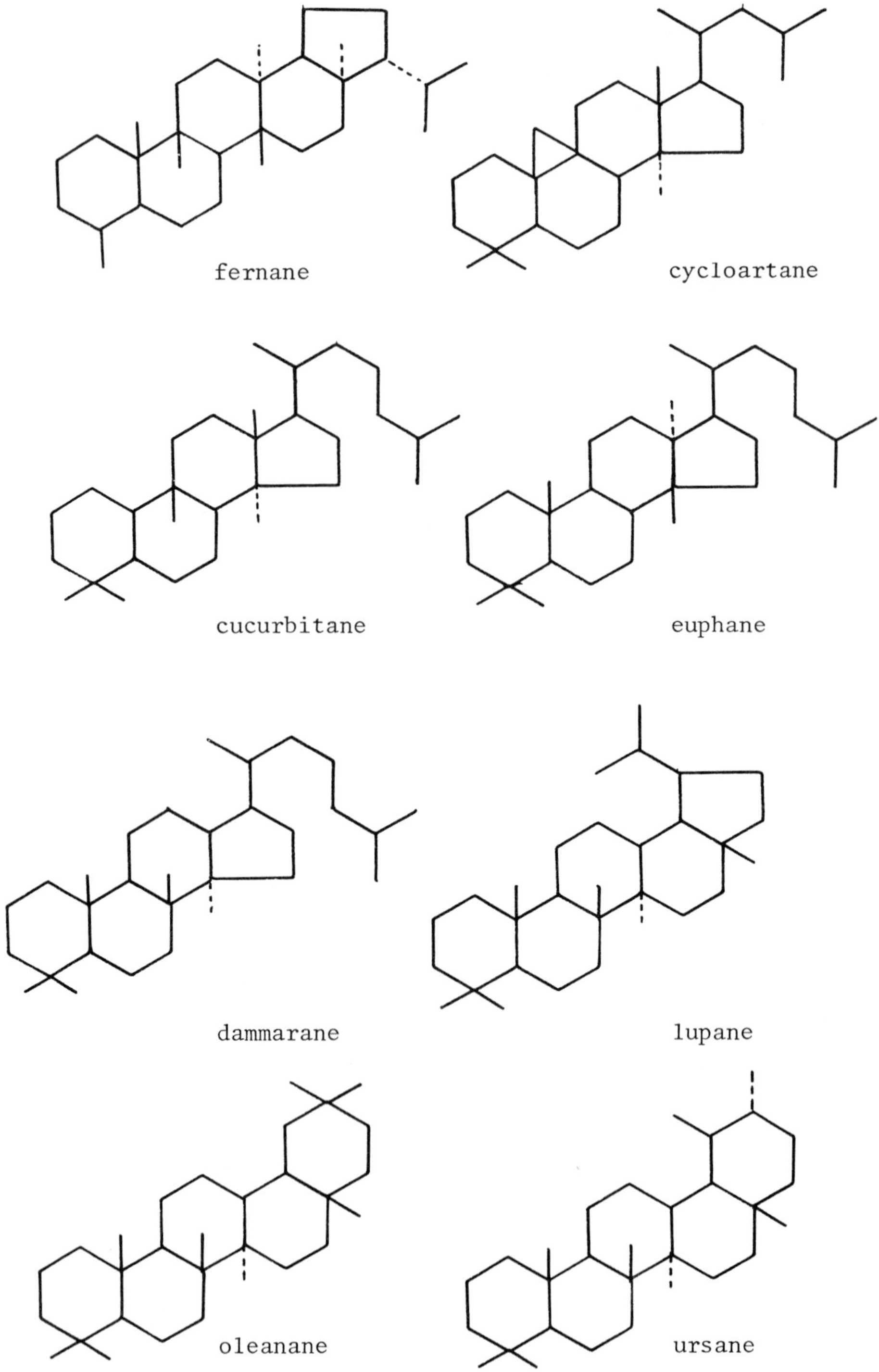

fernane

cycloartane

cucurbitane

euphane

dammarane

lupane

oleanane

ursane

Figure 2: Common Skeletal Types of Triterpenes.

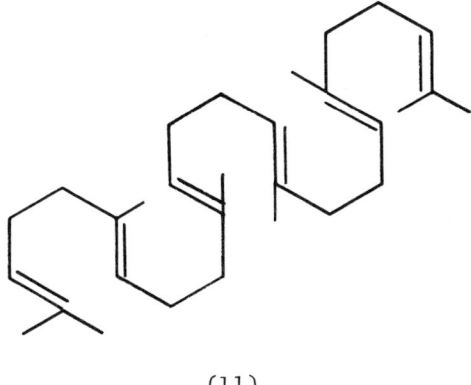

(11)

Fig. 2) are formed in ferns from squalene (in a chair-chair-chair-chair sequence) and not via the intermediacy of squalene-2,3-epoxide (Goodwin 1973). These compounds are characteristic of, but not restricted to, ferns.

When squalene (via the 2,3-epoxide) is cyclized from a chair-boat-chair-boat sequence, the first isolable cyclic intermediate (in plants) is cycloartenol.

Cycloartenol proves to be a key intermediate in the synthesis of certain other triterpenes and especially of plant sterols. Many steroids are common (if not ubiquitous, e.g., β-sitosterol, stigmasterol, and campesterol), and it is probable that this set of pathways exists in all plants (Grunwald 1980; Goad and Goodwin 1972). Steroids are components of membranes in plants.

Buxus alkaloids are derived from cycloartenol and occur only in the Buxaceae.

Cholesterol occurs in small quantities but appears to be a universal constituent of plants. This compound serves as a key intermediate in the synthesis of additional compounds.

PHYTOECDYSONES

Phytoecdysones (such as (14)) occur in *Ajuga* (Lamiaceae), *Achranthes* (Amaranthaceae), *Vitex* (Verbenaceae), *Stachyurus* Stachyuraceae) and in several ferns and gymnosperms (Rees 1971).

(12)

Although complex in appearance, the pathways which lead to this group of compounds are widely distributed, presumably arose early in evolution, and have been maintained in many plant groups.

PROGESTAGENS

Progestagens (such as 12) are steroids with 21 carbon atoms. Among these are pregnenolone and progesterone, which occur as animal hormones but are precursors of plant progestagens as well (Grunwald 1980). Cholesterol is converted to pregnenolone (by C-17 side chain cleavage) in a number of plants although C_{29} sterols (e.g., β-sitosterol may also be converted (Grunwald 1980). Pregnenolone is subsequently converted to progesterone; the reverse reaction also has been observed. A number of other types of steroids are known to be derived from these precursors, among them are the cardenolides or cardiac glycosides.

(13)   (14)

CARDIAC GLYCOSIDES

Cardenolides (such as 14) have 23 carbon atoms and possess a 14-β-hydroxyl and an α,β-unsaturated γ-lactone ring. They are formed by condensation of a C_{21} pregnane derivative (pregnenolone or progesterone) with acetate. The sugar moieties of cardenolides are normal sugars or unbranched aldohexoses mostly of the 6-deoxy or 2-deoxy type. About 50 compounds of this type are known. They are known to occur in the Apocynaceae, Asclepiadaceae, Celastraceae, Brassicaceae, Fabaceae, Euphorbiaceae, Moraceae, Ranunculaceae, Scrophulariaceae, Tiliaceae, Sterculiaceae, and Liliaceae. Related compounds with six membered lactone rings (bufadienolides) occur in the Liliaceae and Ranunculaceae.

CUCURBITACINS

Cucurbitacins have recently been shown to be chemically interrelated to the lanostane series and also products of a chair-boat-chair-boat squalene condensation. These highly oxygenated bitter triterpenes were originally isolated from the Cucurbitaceae but are now known to occur in several other families. Brassicaceae (*Iberis*), Liliaceae (*Phormium*), Begoniaceae (*Begonia*), Scrophulariaceae (*Gratiola*), Rosaceae (*Purshia*) and Primulaceae (*Anagallis*).

SAPOGENINS

Sapogenins are C_{27} steroids which are widely distributed in the Liliaceae, Agavaceae, Amaryllidaceae, and Dioscoreaceae as well as in the Scrophulariaceae and Solanaceae. In plants, sapogenins are combined with sugars to form saponins. Generally, up to five sugar units may form a chain which comprises the glycoside linkage, always β, and always in a C-3 position (Grunwald 1980; Takeda 1972). Cholesterol and β-sitosterol are precursors to the formation of sapogenins. Other steroids such as desmosterol may be involved at times. Other types of steroidal saponins have been reported from the Zygophyllaceae, Fabaceae, Fouquieriaceae, and Arecaceae.

Alkaloids such as solanidine and tomatidine are related to diosgenin in terms of structure and are also derived from cholesterol by unknown routes. These metabolites usually occur as their 3-O-glycosides, e.g., solanin and tomatin (Mann 1978).

Triterpenoid saponins are also widespread in nature (de Mayo 1959; Agarwal and Rastogi 1974).

Squalene also condenses in most plants via a chair-chair-chair-boat sequence to produce a complex series of pentacyclic triterpenes such as α-amyrin, β-amyrin, lupeol, and taraxasterol. These compounds are found in virtually all plants.

Another series of compounds including euphol and tirucallol arise from early intermediates in this sequence and are known to occur in several plant families, especially the Euphorbiaceae and the Rutaceae (Connolly and Overton 1972; Dreyer 1968; Mann 1978).

LIMONOIDS AND QUASSINOIDS

A series of triterpenes derived from apoeuphol and euphol (15) (or tirucallol) is widespread in the Rutaceae, Meliaceae, and Simaroubaceae. Of these compounds, the limonoids are the simplest; others such as the quassinoids and B and C ring opened compounds are derived from them (Connolly, et al. 1970; Dreyer 1968; Dreyer, et al. 1972; and Polonsky 1973). The presence of several intermediates such as flindissol (16) (Rutaceae), turreanthin (Meliaceae), and melianodiol (Simaroubaceae) suggest the mode of formation of the furan ring of limonoids. As a second major step in the synthesis of both limonoids and quassinoids, four terminal atoms of the side chain are lost and the residue converted into a furan ring. The rearrangement from

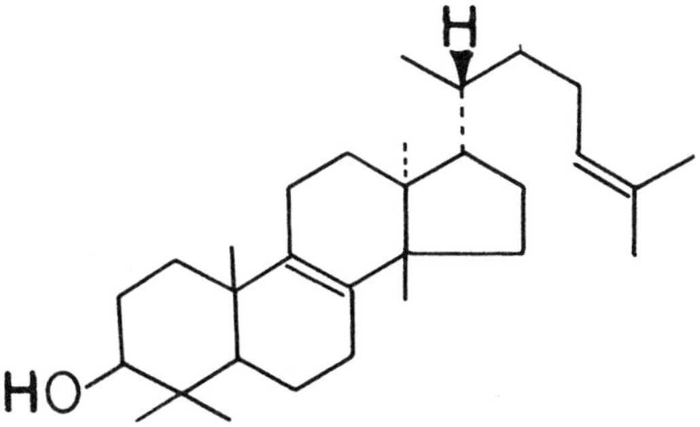

euphol (15)

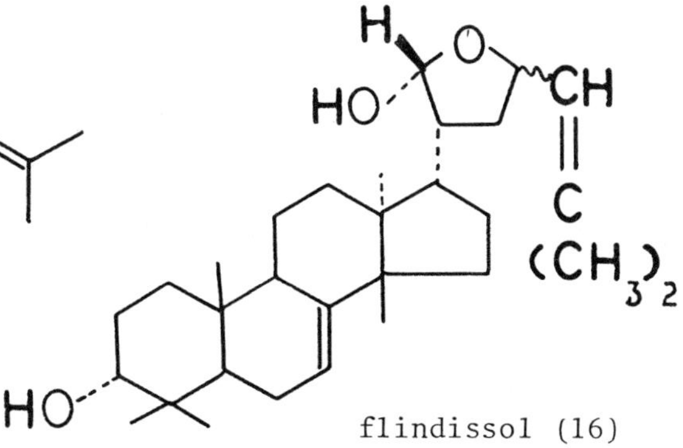

flindissol (16)

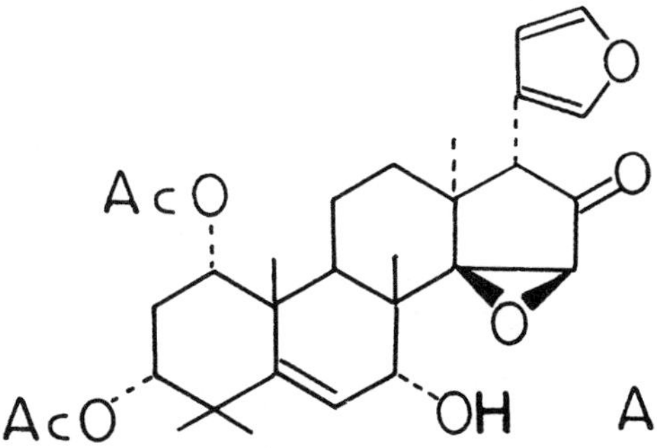

grandifolione (17)

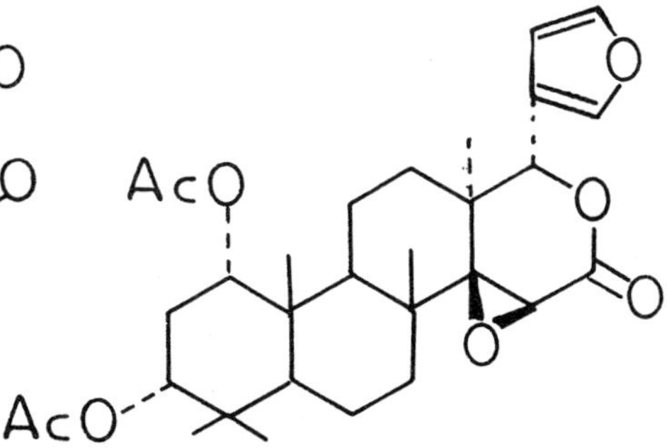

khivorin (18)

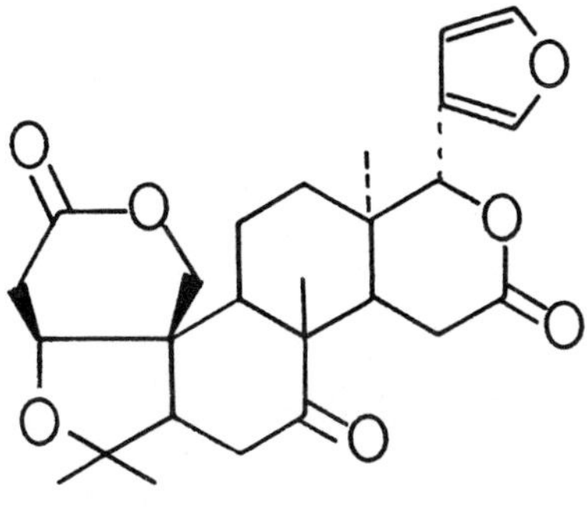

limonin (19)

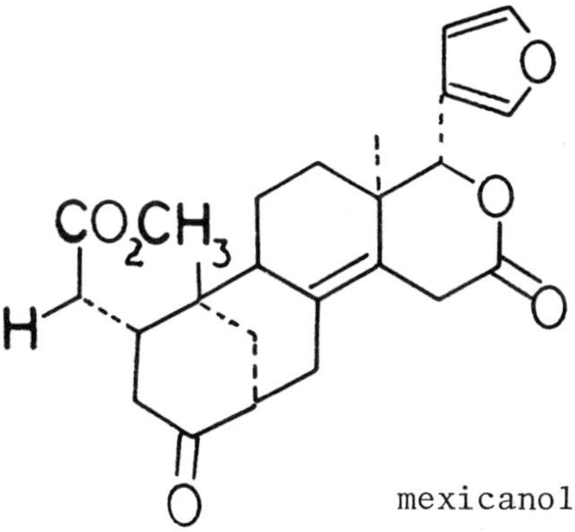

mexicanolide (20)

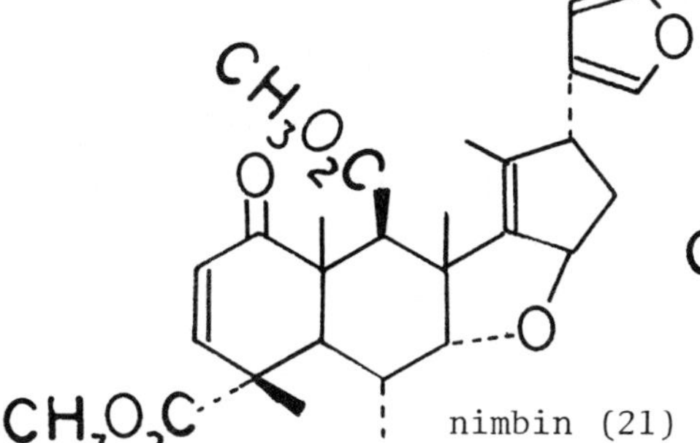

nimbin (21)

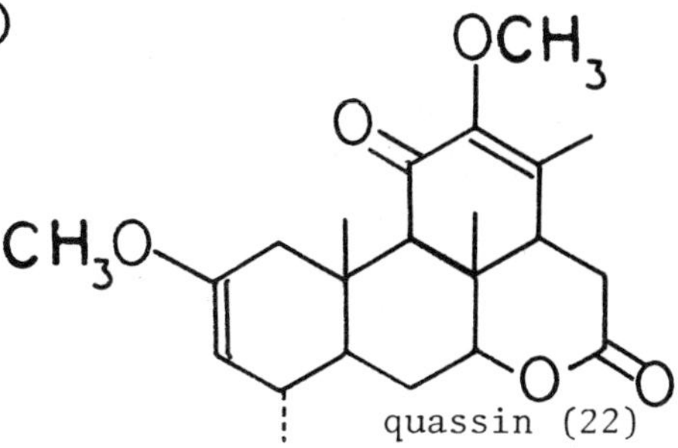

quassin (22)

(23)

euphol to apo-euphol has occurred. By a series of subsequent reactions, compounds such as grandifolione (17) and khivorin (18) are formed, which possess epoxidized and additionally oxidized D-ring structures.

In the Rutaceae, compounds of this type are cleaved in the A-ring to produce limonin (19) and a series of related compounds. In the Meliaceae, B- and C-ring cleavage is often observed to yield compounds such as mexicanolide (20) and nimbin (21), respectively. In the Simaroubaceae, a series of C_{20} compounds (quassinoids, e.g., 22) arise from limonoid precursors. Simarolide (23) (a C_{25} compound) is intermediate in many regards between the two groups. In the formation of these compounds, methyl groups and the furan ring found in limonoids are lost. The oxygenated D-ring is opened and recyclized. Several compounds which occur in the Rutaceae and Cneorideae appear to be derived from limonoids, but little work has been done on their biogenesis.

This series of compounds and pathways provides one of the best examples of the application of terpene chemistry to phylogeny. From the above considerations, it is possible to conclude that the Meliaceae and Simaroubaceae arose from Rutaceae-like ancestors. The Simaroubaceae has the most highly modified chemistry of the three families. Within the Rutaceae, chemical changes (synthesis of limonin) have occurred which presumably arose after the separation of the Meliaceae and the Simaroubaceae from the Rutaceous stock.

TETRATERPENES

The properties and structures of tetraterpenes differ markedly from those of the lower terpenes (Ramage 1972). The carotenoids (the major representative of this class) are mostly linear polyene systems. These compounds often possess large numbers of possible geometric isomers, but in practice most double bonds exist in the more stable trans form. The high degree of unsaturation renders them both heat and light sensitive and they are an experimentally demanding group of compounds with which to work (Ramage 1972). The leaves of all green plants contain the same major carotenoids: β-carotene, lutein, violaxanthin, and neoxanthin. Photosynthetic carotenoids are thought to occur as carotenoid-protein complexes in the living plant. Carotenoids are also found in fruits (e.g., lycopene in tomatoes and capsanthin in red pepper) and flowers (which often have β-carotene or lycopene).

The general aspects of carotenoid biosynthesis have been reviewed (Britton 1971). Although carotenoids have been of limited taxonomic value in higher plants, they have proven useful in studies of algae and bacteria (Britton 1971; Goodwin 1971).

LITERATURE CITED

Adams, R. P. 1977. Chemosystematics-analysis of populational differentiation and variability of ancestral and recent populations of *Juniperus ashei*. Ann. Mo. Bot. Gard. 64: 184-209.

Agarwal, S. K. and R. P. Rastogi. 1974. Triterpenoid saponins and their genins. Phytochem. 13:2623-2645.

Alston, R. E. and B. L. Turner. 1963. Biochemical Systematics. Englewood Cliffs, New Jersey: Prentice-Hall.

Banthorpe, D. V. and B. V. Charlwood. 1980. "The Terpenoids". In Secondary Plant Products, edited by E. A. Bell and B. V. Charlwood, pp. 185-220. Berlin: Springer Verlag.

Banthorpe, D. V., B. V. Charlwood and M. J. O. Francis. 1972. The biosynthesis of monoterpenes. Chem. Rev. 72:115-156.

Bate-Smith, E. C., I. K. Ferguson, K. Hutson, S. R. Jensen, B. J. Nielsen and T. Swain. Phytochemical interrelationships in the Cornaceae. Biochem. Syst. & Ecol. 3:79-89.

Birch, A. J. 1974. "Biosynthetic Pathways in Chemical Phylogeny." In Chemistry in Botanical Classification, edited by G. Bendz and J. Santesson, pp. 261-270. New York: Academic Press.

Britton, G. 1971. "General Aspects of Carotenoid Biosynthesis". In *Aspects of Terpenoid Chemistry and Biochemistry*, edited by T. W. Goodwin, pp. 255-289. London: Academic Press.

Charlwood, B. V. and D. V. Banthorpe. 1978. "The Biosynthesis of Monoterpenes". In *Progress in Phytochemistry*, Vol. 5, edited by L. Reinhold, J. B. Harborne, and T. Swain, pp. 62-125. Oxford: Pergamon Press.

Coates, R. M. 1976. Biogenetic type rearrangements of terpenes. *Fort. Chem. Org. Naturst.* 33:73-230.

Connolly, J. D. and K. H. Overton. 1972. "The Triterpenoids." In *Chemistry of Terpenes and Terpenoids*, edited by A. A. Newman, pp. 207-287. London: Academic Press.

Connolly, J. D., K. H. Overton and J. Polonsky. 1970. "The Chemistry and Biochemistry of the Limonoids and Quassinoids." In *Progress in Phytochemistry*, Vol. 2, edited by B. Reinhold and Y. Liwschitz, pp. 385-455. London: Interscience.

Cordell, G. A. 1976. Biosynthesis of sesquiterpenes. *Chem. Revs.* 76:425-460.

------. 1974. The biosynthesis of indole alkaloids. *Lloydia* 37:219-298.

Coscia, C. J. 1969. "Picrotoxins." In *Cyclopentanoid Terpene Derivatives*, edited by W. I. Taylor and A. R. Battersby, pp. 147-201. New York: Dekker.

Cronquist, A. 1968. *The Evolution and Classification of Flowering Plants*. Boston: Houghton Mifflin.

Dahlgren, R. 1977. A commentary on a diagrammatic presentation of the angiosperms in relation to the distribution of character states. *Plant Syst. Evol. Suppl.* 1:253-283.

------. 1975. A system of classification of the angiosperms to be used to demonstrate the distribution of characters. *Bot. Not.* 128:119-147.

Dahlgren, R., S. R. Janzen and B. J. Neilsen. 1976. Iridoid compounds in Fouquieriaceae and notes on its possible affinities. *Bot. Notiser.* 129:207-212.

Davis, P. and V. H. Heywood. 1963. Principles of Angiosperm Taxonomy. Princeton, New Jersey: Von Nostrand Reinhold.

De Candolle, A. P. 1804. Essai sur les propiétés médicales des plantes, compareés avec leurs formes extérieures et leur classification naturelle. 1st edition, Paris. (Second edition, Paris, 1816).

de Mayo, P. 1959. The Higher Terpenoids. New York: Interscience.

Devon, T. K. and A. I. Scott. 1972. Handbook of Naturally Occurring Compounds, Vol. 2. New York: Academic Press.

Dreyer, D. L. 1968. Limonoid bitter principles. Fort. Chem. Org. Naturst. 26:190-244.

Dreyer, D. L., M. V. Pickering and P. Cohan. 1972. Distribution of limonoids in the Rutaceae. Phytochem. 11:705-713.

Fairbrothers, D. E., T. J. Mabry, R. L. Scogin and B. L. Turner. 1975. The bases of angiosperm phylogeny: chemotaxonomy. Ann. Mo. Bot. Gard. 62:765-800.

Fischer, N. H. 1978. On the biogenesis of pseudoguaianolides. Rev. Latinoamer. Quim. 9:41-46.

Fischer, N. H., E. J. Olivier and H. D. Fischer. The biogenesis and chemistry of sesquiterpene lactones. Fort. Chem. Org. Naturst. 38:47-388.

Francis, M. J. O. 1971. "Monoterpene Biosynthesis." In Aspects of Terpenoid Chemistry and Biochemistry, edited by T. W. Goodwin, pp. 29-51. London: Academic Press.

Geissman, T. A. and D. H. G. Crout. 1969. Organic Chemistry of Secondary Plant Metabolism. San Francisco: Freeman Cooper.

Gibbs, R. D. 1974. Chemotaxonomy of Flowering Plants. Montreal: McGill-Queens University Press.

Goad, L. J. and T. W. Goodwin. 1972. "The Biosynthesis of Plant Sterols." In Progress in Phytochemistry, Vol. 3, edited by L. Reinhold and Y. Liwschitz, pp. 113-198. London: Interscience.

Goodwin, T. W. 1973. "Recent Developments in the Biosynthesis of Plant Triterpenes." In Terpenoids: Structure, Biogenesis and Distribution, (Rec. Adv. Phytochem. Vol. 6), edited by V. C. Runeckles and T. J. Mabry, pp. 97-116. New York: Academic Press.

------. 1971. "Algal Carotenoids." In Aspects of Terpenoid Chemistry and Biochemistry, edited by T. W. Goodwin, pp. 315-356. London: Academic Press.

Greshoff, M. 1893. Gedanken über Pflanzenkräfte und phytochemische Verwandtschaft. Ber. Deut. Pharm. Ges. 3:191-204.

Grunwald, C. 1980. "Steroids." In Secondary Plant Products, edited by E. A. Bell and B. V. Charlwood, pp. 221-256. Berlin: Springer Verlag.

Guenther, E. 1948-1952. The Essential Oils, Vols. 1-6. New York: Van Nostrand.

Guenther, E., G. Gilbertson and R. T. Koenig. 1977. Essential oils and related products. Anal. Chem. 49:83R-98R.

Gupta, B. O., S. K. Bannerjee and K. L. Handa. 1976. Alkaloids and coumarins of *Heracleum wallichii*. Phytochem. 15:576.

Hanson, J. R. 1972. "The Bicyclic Diterpenes." In Progress in Phytochemistry, Vol. 3, edited by L. Reinhold and Y. Liwschitz, pp. 231-285. London: Interscience.

------. 1971. The biosynthesis of the diterpenes. Fort. Chem. Org. Natur. 29:395-416.

Harborne, J. B. 1977. Introduction to Ecological Biochemistry. London: Academic Press.

Hecker, E. 1971. "Cocarcinogens from Euphorbiaceae and Thymeliaceae." In Pharmacognosy and Phytochemistry, edited by H. Wagner and L. Hörhammer, pp. 147-165. Berlin: Springer Verlag.

Hegnauer, R. 1978. The importance of essential oils in plant classification. Dragoco Report:203-230.

------. 1976. "Accumulation of Secondary Products and its Significance for Biological Systems." In Secondary Metabolism and Coevolution, edited by M. Luckner, K. Mothes and

L. Nover, pp. 45-66. Halle: Deutsche Akad. Naturf. Leopoldina.

------. 1969. "Chemical Evidence for the Classification of Some Plant Taxa." In Perspectives in Phytochemistry, edited by J. B. Harborne and T. Swain, pp. 121-138. London: Academic Press.

------. 1967. Chemical characters in plant taxonomy: some possibilities and limitations. Pure & Appl. Chem. 14:173-187.

------. 1962-1973. Chemotaxonomie der Pflanzen, Vol. 1-6 Basel: Birkhauser Verlag.

------ and P. Kooiman. 1978. Die systematische bedeutung von iridoiden Inhaltstoffen im Rahmen von Wettstein's Tubiflorae. Plant Med. 33:1-33.

Herout, V. 1973. "Biochemistry of Sesquiterpenes". In Aspects of Terpenoid Chemistry and Biochemistry, edited by T. W. Goodwin, pp. 53-94. New York: Academic Press.

Heywood, V. H. 1971. The Biology and Chemistry of the Umbelliferae. New York: Academic Press.

------, J. B. Harborne and B. L. Turner. 1977. The Biology and Chemistry of the Compositae, Vols. 1 & 2. London: Academic Press.

Irving, R. S. and R. P. Adams. 1973. "Genetic and Biosynthetic Relationships of Monoterpenes." In Terpenoids: Structure, Biogenesis and Distribution (Rec. Adv. Phytochem., Vol. 6), edited by V. C. Runeckles and T. J. Mabry, pp. 187-214. New York: Academic Press.

Karrer, W., E. Cherbuliez and C. H. Eugster. 1977. Konstitution und Vorkommen der Organischen Pflanzenstoffe (exclusive Alkaloids), Erganzungsband 1, Basel: Birkhäuser Verlag.

Kinghorn, A. D. 1979. "Cocarcinogenic Irritant Euphobiaceae." In Toxic Plants, edited by A. D. Kinghorn, pp. 137-159. New York: Columbia University Press.

Kubitzki, K. 1969. Chemosystematische Betrachtungen zur Grossgliederung der Dicotylen. Taxon 18:360-368.

Luckner, M. 1972. Secondary Metabolism in Plants and Animals. New York: Academic Press.

Mabry, T. J. 1974. "The Chemistry of Disjunct Taxa." In Chemistry and Botanical Classification, edited by G. Bendz and J. Santesson, pp. 63-66. New York: Academic Press.

MacMillan, J. 1971. "Diterpenes--The Gibberellins." In Aspects of Terpenoid Chemistry and Biochemistry, edited by T. W. Goodwin, pp. 153-180. New York: Academic Press.

Mann, J. 1978. Secondary Metabolism. Oxford: Oxford University Press.

McNair, J. B. 1965. Studies in Plant Chemistry, Including Chemical Taxonomy, Ontogeny, Phylogeny, etc. Los Angeles: publ. by the author.

Meeuse, A. D. J. 1970. The descent of the flowering plants in the light of new evidence from phytochemistry and from other sources. II. Suggestions for a holotaxonomic major classification. Acta Bot. Neerl. 19:133-140.

Mothes, K. 1966. Biogenesis of alkaloids and the problem of chemotaxonomy. Lloydia 29:156-171.

Nakanishi, K., T. Goto, S. Ito, S. Natori and S. Nozoe. 1974. Natural Products Chemistry, Vol. 1. New York: Academic Press.

Newman, A. A. 1972. Chemistry of Terpenes and Terpenoids. London: Academic Press.

Oehlschlager, A. C. and G. Ourisson, 1967. "A Comparison of in vivo and in vitro Skeletal Transformations of Diterpenes". In Terpenoids in Plants, edited by J. B. Pridham, pp. 83-110. London: Academic Press.

Pant, P. and R. P. Rastogi. 1979. The triterpenoids. Phytochem. 18:1095-1108.

Polonsky, J. 1973. "Chemistry and Biogenesis of the Quassinoids (Simaroubolides)". In Terpenoids: Structure, Biogenesis, and Distribution, (Rec. Adv. Phytochem., Vol. 6), edited by V. C. Runeckles and T. J. Mabry, pp. 31-64. New York: Academic Press.

Radford, A. E., W. C. Dickinson, J. R. Massey and C. R. Bell. 1974. Vascular Plant Systematics. New York: Harper.

Ramage, A. 1972. "Carotenoid Chemistry". In *Chemistry of Terpenes and Terpenoids*, edited by A. A. Newman, pp. 288-336. London: Academic Press.

Rees, H. H. 1971. "Ecdysones". In *Aspects of Terpenoid Chemistry and Biochemistry*, edited by T. W. Goodwin, pp. 181-222. New York: Academic Press.

Rosenthal, G. A. and D. H. Janzen. 1979. *Herbivores*. New York: Academic Press.

Rücker, G. 1973. Sesquiterpenes. *Angew. Chem. Int. Ed.* 12:793-806.

Seigler, D. 1981 (in press). "Secondary Metabolites and Plant Systematics." In *The Biochemistry of Plants, Secondary Plant Products*, Vol. 7, edited by E. E. Conn. New York: Academic Press.

------. 1977. "Plant Systematics and Alkaloids." In *The Alkaloids*, Vol. 16, edited by R. H. F. Manske, pp. 1-82. New York: Academic Press.

------. 1974. Chemists and taxonomy. *Chem. in Brit.* 10:339-342.

Stebbins, G. L. 1974. *Flowering Plants*. Cambridge, Massachusetts: Belknap Press.

Takeda, K. 1972. "The Steroidal Sapogenins of the Dioscoreaceae." In *Progress in Phytochemistry*, edited by L. Reinhold and Y. Liwschitz, pp. 287-333. London: Interscience.

Taylor, W. I. and A. R. Battersby. 1969. *Cyclopentanoid Terpene Derivatives*. New York: Dekker.

Tétényi, P. 1970. *Infraspecific Chemical Taxa of Medicinal Plants*. New York: Chemical Publ. Co.

Thorne. 1976. "A Phylogenetic Classification of the Angiospermae". In *Evolutionary Biology*, Vol. 9, edited by M. K. Hecht, W. C Steere and B. Wallace, pp. 35-106. New York: Plenum Press.

von Rudloff, E. 1975. Volatile leaf oil analysis in chemosystematic studies of North American conifers. *Biochem. Syst. & Ecol.* 2:131-167.

von Rudloff, E. 1969. "Scope and Limitations of Gas Chromatography of Terpenes in Chemosytematic Studies." In Rec. Adv. Phytochem., Vol. 2, edited by M. K. Seikel and V. C. Runeckles, pp. 127-162. New York: Appleton Century Crofts.

Yoshioka, H., T. J. Mabry and B. N. Timmermann. 1973. Sesquiterpene Lactones. Tokyo: University of Tokyo Press.

Young, D. A. and D. S. Seigler. 1981 (in press). "General Classification and Characteristics of Vascular Seed Plants." In Basic Principles, Vol. 1, Biosolar Resources, edited by A. Mitsui and C. C. Black. West Palm Beach, Florida: CRC Publishing Co.

A REVISED CLASSIFICATION OF THE ANGIOSPERMS
WITH COMMENTS ON CORRELATION BETWEEN CHEMICAL
AND OTHER CHARACTERS

Rolf M. T. Dahlgren

Botanical Museum of
the University of Copenhagen
Gothersgade 130
DK 1123 Copenhagen
DENMARK

*Søren Rosendal-Jensen and
Bent Juhl Nielsen*

Institute of Organic Chemistry
The Technical University
DK 2800 Lyngby
DENMARK

It is beginning generally to be acknowledged, and has been proved very effectively at this symposium, that chemical properties can be used to a wide extent in taxonomy. It is also equally obvious that it is not the presence or absence (in themselves very relative concepts) of particular compounds that is of importance, but the biosynthetic pathways of the compounds. The same compound thus may have originated along totally or partly different routes, and even when formed along the same biosynthetic pathway, the pathway may have evolved independently in different evolutionary lines. The crucial questions are then, when and where can chemical characters be used with justification in taxonomy? That is, when and where do they indicate phylogenetic relationships? The best indication may be when chemical characters show a convincing correlation with other independent attributes, which may be in the fields of morphology, anatomy, embryology, cytology, pollen morphology, etc.

A revised system of classification of the angiosperms and a new diagram of their orders is presented below, in the framework of which the distribution of some chemical characters are

Figure 1. Superorders of angiosperms shown as clusters of orders in a two-dimensional framework (names of the orders are given in Figures 2-8).

inserted. Chemical characters are considered along with the distribution of other characters. Where possible, examples of families will be given which previously have been misplaced, and the affinities of which have become increasingly clear on the basis of chemical characters used in combination with other, especially embryological, features. Examples will be given for the following three categories: (1) the angiosperms at large; (2) groups synthesizing iridoid compounds; and (3) the monocotyledons.

CLASSIFICATION OF THE ANGIOSPERMS

A provisional classification of the angiosperm families and orders was given in a previous paper (Dahlgren 1975a) in connection with the presentation of a diagram showing the orders dimensioned after their size and placed according to their presumed mutual affinities. The main purpose of this classification and diagram was to provide a framework for illustrating the distribution of supposedly important character states. Since then a number of attributes have been plotted on this diagram (e.g., Dahlgren 1975b, 1977a, b, 1979; Behnke and Dahlgren 1976, 1979; Gornall, et al. 1979). An intention of this project was naturally also to achieve a basis for an improved classification of the angiosperms, showing a higher degree of predictability with regard to concealed or unconcealed characters. The predictive value of a well-balanced classification is perhaps the best reason for spending time and effort in appreciating phylogenetic relationships among families and orders of plants, and in doing so a two-dimensional diagram is more helpful than a list.

A revised classification is presented here (Appendix 1) which is based upon results from the above projects investigating the distribution of attributes which are presumed to express phylogenetic relationships (though they are generally not monophyletic). Recently published results have been considered as far as possible. In making this classification, there have been more base data for the monocotyledons than for the dicotyledons, as the former have been studied intensively with regard to the distribution of a great number of character states by Dahlgren and Clifford in the last two years. The classification and diagram (Fig. 1) are largely self explanatory, but it must again be stressed that a diagram such as this cannot fulfil all requirements for expressing the desired alignments. The intent that closely allied orders should be placed in close approximity to each other cannot always be attained in a two-dimensional pattern, especially when orders largely have been dimensioned based upon the number of species; these requirements are often partly incompatible. The known degree of affinity in characters very rarely can be expressed as distances in a diagram, and even when the affinity can be measured for a few taxa (e.g., a serological test), affinities can rarely be estimated in terms of distance when several large groups are compared; the task is totally impossible when numerous attributes are considered. Convergent evolution in unrelated groups of plants also may result in surprising similarities, and is difficult to uncover when insufficient data are available. An example of a defect caused by spatial problems in the present diagram (Fig. 1) can be mentioned. The trochodendralean end of the Hamamelidiflorae is not connected as closely to the Magnoliiflorae as would have been justified from its similarity in apocarpy, cellular endosperm formation, primitive vascular elements, etc., attributes generally agreed to be primitive character states. On the other hand, the spacial position in the diagram of the Trochodendrales makes possible a number of other connections.

It needs to be pointed out that the classification presented here (see Appendix 1) is very different in its appearance from that presented in Dahlgren (1975). However, the differences are largely superficial, because within the present diagram many orders and superorders forming supposedly allied aggregates are connected as in the 1975 diagram, and their successive positions in relation to each other are partly comparable. Some orders with very uncertain affinities, like the Polygonales and Plumbaginales, have positions different from those in the diagram of 1975. For other groups, new evidence indicates new alignments. Thus, the Nymphaeales and Piperales have been placed closer together, as have the Dipsacales and Cornales. It is still with some doubt that Myricales and Juglandales have been placed next to the Fagales, since they also show great affinity to the sapindalean Anacardiaceae. These realignments have been based upon new unpublished serological results (Fairbrothers and Petersen, pers. com) as well as similarity in pollen morphology, chalazogamy, tannins, etc., although the number of anacardiaceous features in Juglandaceae and Myricaceae also are striking (Thorne 1973). The Tropaeolales has been erected and placed near the Geraniales, the "classical" position (see discussion below under Glucosinolates).

Flavonoid chemistry (Gornall, et al. 1979), seed coat struc-

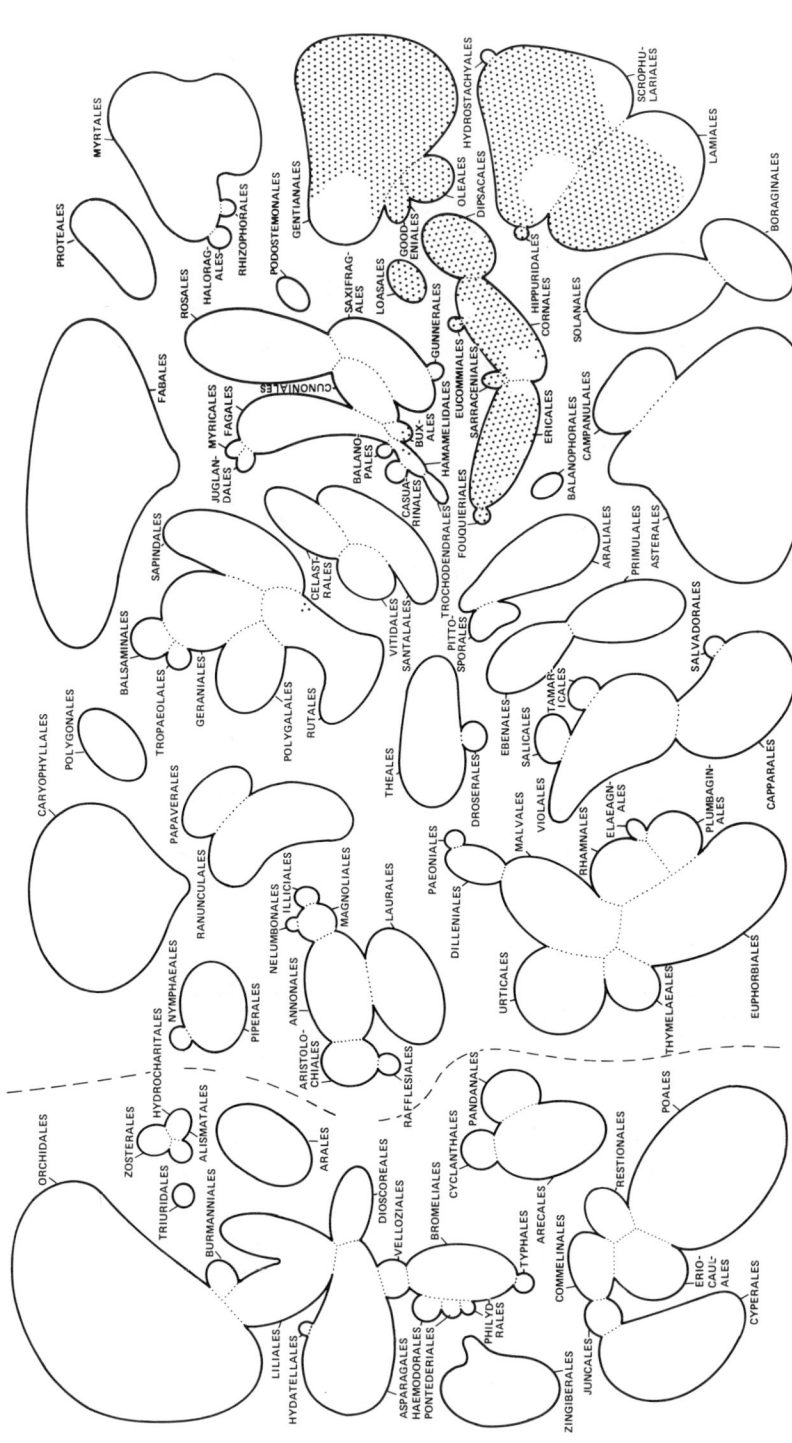

Figure 2. Approximate distribution of iridoids in the angiosperms (details are given in the text).

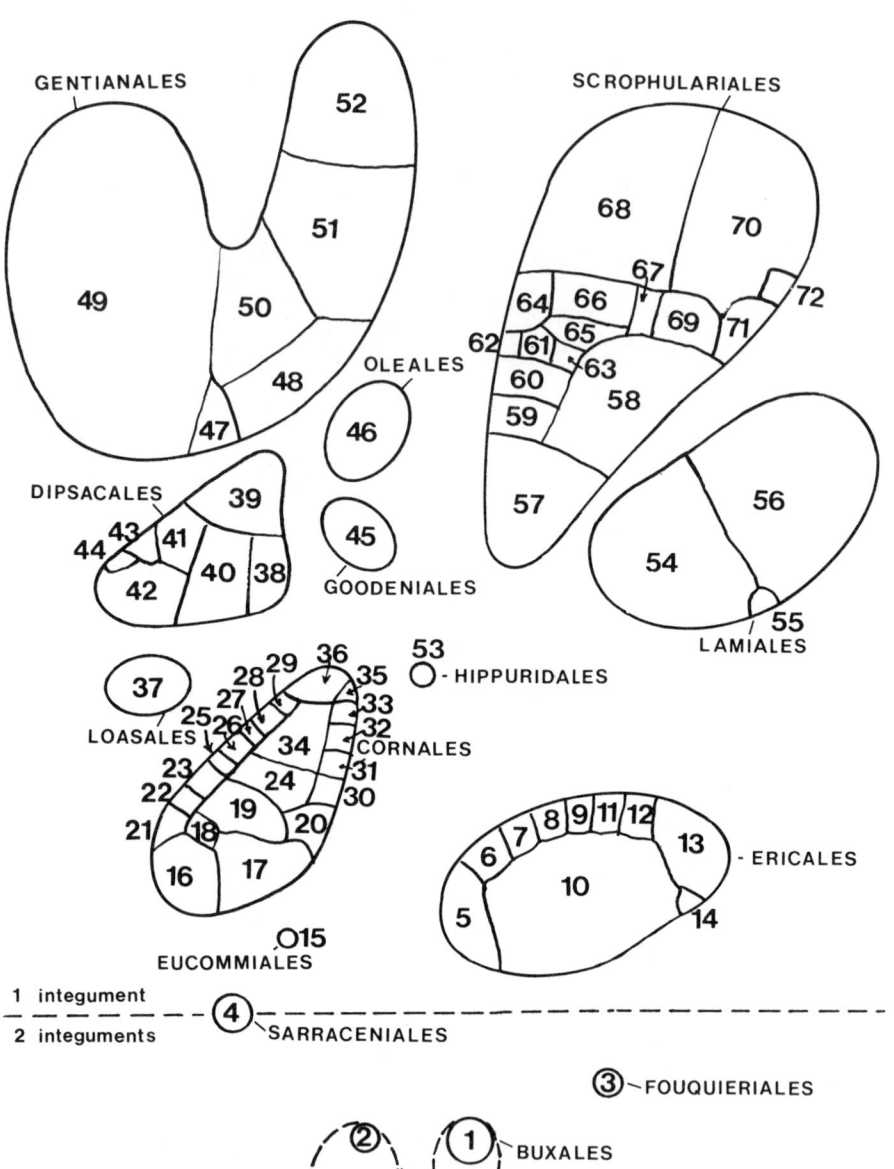

Figure 3. Orders of angiosperms containing iridoids presented separately (numbers refer to the families in Table 2).

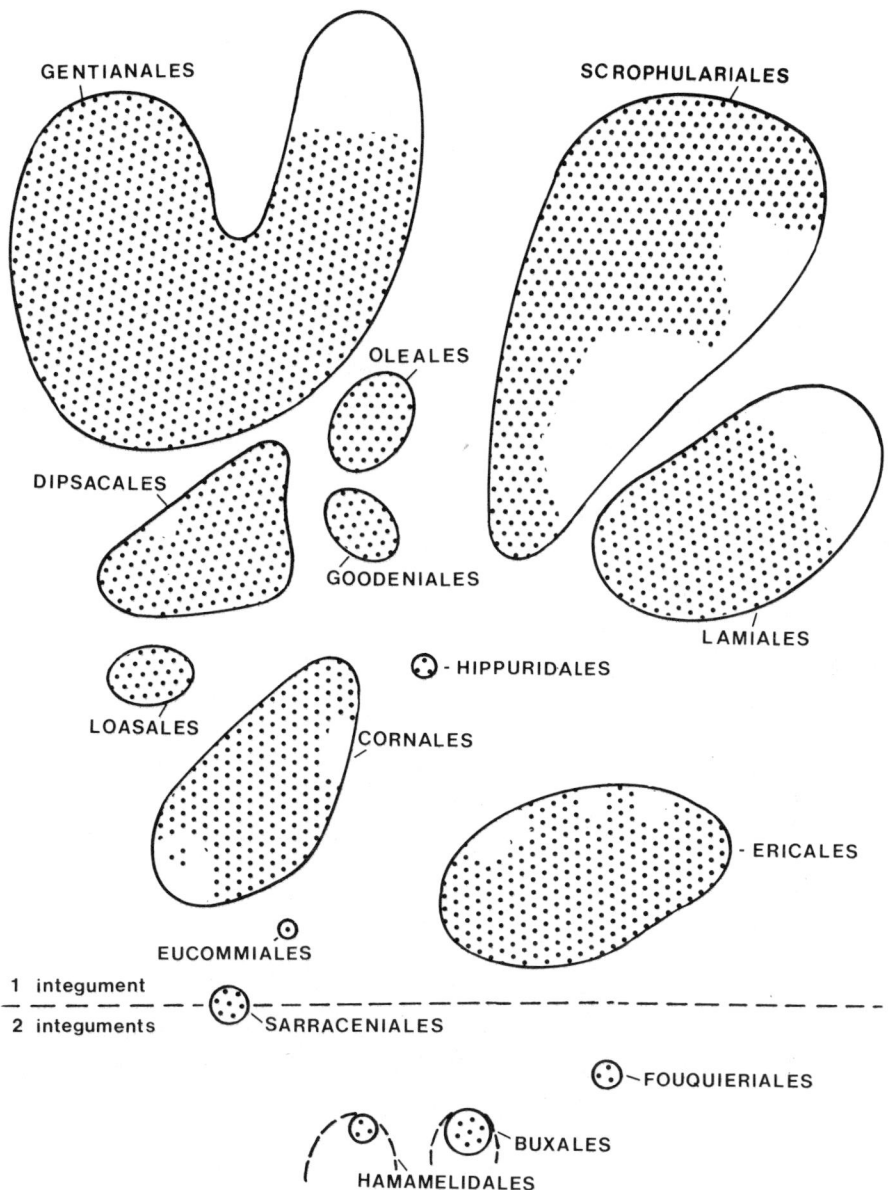

Figure 4. Iridoid-containing families (shaded). Non-iridoid parts of Lamiaceae, Bigoniaceae and Acanthaceae left unshaded (Gesneriaceae is not yet known to contain iridoids); otherwise the families are shaded provided iridoids are known in any member (cf. Fig. 3).

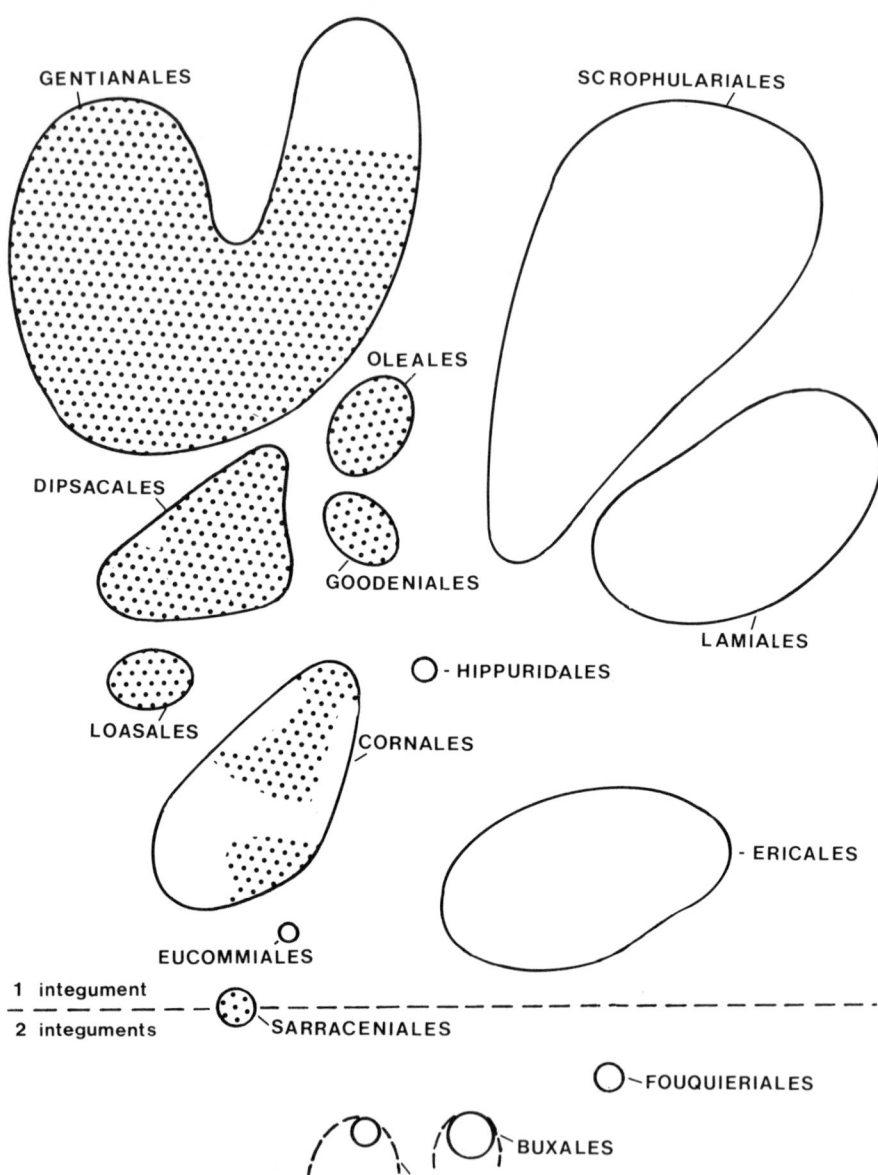

Figure 5. Families containing seco-iridoids (cf. Fig. 3).

Table 1: Classification of iridoid compounds according to Jensen, et al. (1975).

Carboxylic Iridoids

Group I. 10-hydroxylated compounds.
Group II. 8 b-Oxy-8-methyl compounds.
Group III. Cornin group.
Group IV. 10-carboxyl and 10-decarboxylated iridoids.
Group V. Valeriana compounds.

Seco-iridoids

Group VI. Simple secoiridoids.
Group VII. Gentiopicroside groups.
Group VIII. Oleuropein group.
Group IX Complex iridoid alkaloids.

Primitive or Otherwise Non-Classified Alkaloids

Group X.

Table 2. Iridoid-containing orders and families. An asterisk (*) denotes families where iridoids have been identified.

Rutales: *Meliaceae (only *Xylocarpus*) (not shown in Figs. 3-6), Rutaceae, etc.

Buxales: (Daphniphyllaceae (*Daphniphyllum*) (1), Buxaceae

Hamamelidales: *Hamamelidaceae (*Liquidambar*) (2), Platanaceae, etc.

Fouquieriales: *Fouquieriaceae (3)

Sarraceniales: *Sarraceniaceae (4)

Ericales: *Actinidiaceae (5), Clethraceae (6), Diapensiaceae (7), Pyrolaceae (8), Empetraceae (9), *Ericaceae (10), *Monotropaceae (11), Cyrillaceae (12), *Epacridaceae (13), *Roridulaceae (14)

Eucommiales: *Eucommiaceae (15)

Cornales: *Symplocaceae (16), *Icacinaceae (17), *Montiniaceae (18), *Escalloniaceae (19), *Stylidiaceae (20), *Alangiaceae (21), *Nyssaceae (22), *Garryaceae (23), *Cornaceae (incl. *Corokia*) (24), *Davidiaceae (25), *Mastixiaceae (26), Helwingiaceae (27), *Griselinaceae (28), *Aucubaceae (29), *Curtisiaceae (30), (*Aralidiaceae not in the diagram), Alseuosmiaceae (21), Columelliaceae (32), *Torricelliaceae (33), *Hydrangeaceae (34), *Adoxaceae (35), *Sambucaceae (36). Anisophyllaceae, Aquifoliaceae, Dialypetalanthaceae Dulongiaceae, Eremosynaceae, Medusandraceae, Paracryphiaceae, Phellinaceae, Pterostemonaceae, Sphenostemonaceae, Tetracarpaeaceae and Tribelaceae from a morphological point of view are likely to belong in Cornales, but have been omitted here for the sake of simplicity. Hardly any of them has been investigated for iridoids.

Loasales: *Loasaceae (37)

Dipsacales: *Viburnaceae (38), *Valerianaceae (39), *Caprifoliaceae (40), *Calyceraceae (41), *Dipsaceaeae (42), Morinaceae (43), Triplostegiaceae (44)

Goodeniales: *Goodeniaceae (45)

Oleales *Oleaceae (46)

Gentianales: *Menyanthaceae (47), *Loganiaceae (48), *Rubiaceae (49), *Gentianaceae (50), *Apocynaceae (51), Asclepiadaceae (52)

Hippuridales: *Hippuridaceae (53)

Lamiales: *Verbenaceae (54), *Callitrichaceae (55), *Lamiaceae (56)

Scrophulariales: *Bignoniaceae (57), Gesneriaceae (58), *Myoporaceae (59), *Buddlejaceae (60), *Stilbaceae (61), *Retziaceae (62), *Globulariaceae (63), *Selaginaceae (64), *Pedaliaceae (65), *Plantaginaceae (66), *Martyniaceae (67), *Scrophulariaceae (incl. Orobanchaceae) (68), *Lentibulariaceae (69), *Acanthaceae (70), Thunbergiaceae (71), Mendonciaceae (72)

Table 3. Occurrence of iridoid groups within the angiosperms. (non-iridoid producing families are omitted.)

Order	Family	I	II	III	IV	V	VI	VII	VIII	IX	X
Rutales	Meliaceae						+				
Buxales	Daphniphyllaceae	+									
Hamamelidales	Altingiaceae	+									
Fouquieriales	Fouquieriaceae	+								+	
Sarraceniales	Sarraceniaceae						+				
Ericales	Actinidiaceae										+
	Pyrolaceae	+									
	Ericaceae	+			+						
	Monotropaceae	+									
	Epacridaceae										+
	Roridulaceae										+
Eucommiales	Eucommiaceae	+	+								
Cornales	Symplocaceae			+							
	Icacinaceae	+					+			+	+
	Montiniaceae									+	
	Escalloniaceae	+									
	Stylidiaceae	+					+				
	Alangiaceae									+	
	Nyssaceae									+	
	Garryaceae	+									
	Cornaceae s.str.	+		+							
	Davidiaceae						+				+
	Mastixiaceae										+
	Curtisiaceae						+				
	Griselinaceae			+							
	Torricelliaceae			+							
	Aucubaceae	+									
	Aralidiaceae			+							
	Hydrangeaceae	+			+	+	+				+
	Adoxaceae						+				+
	Sambucaceae						+				
Loasales	Loasaceae					+	+	+			+
Dipsacales	Viburnaceae			+		+	+	+			+
	Valerianaceae					+	+				+
	Caprifoliaceae						+				+
	Calyceraceae						+				
	Dipsacaceae						+	+			+
Goodeniales	Goodeniaceae						+				+
Oleales	Oleaceae					+			+	+	+
Gentianales	Menyanthaceae						+				+

Table 3 (continued).

		1	2	3	4	5	6	7
	Loganiaceae				+		+	+
	Rubiaceae	+	+		+		+	+
	Gentianaceae				+	+		+
	Apocynaceae	+			+	+	+	+
Hippuridales	Hippuridaceae	+						
Lamiales	Verbenaceae	+	+	+	+			+
	Callitrichaceae	+						
	Lamiaceae	+	+					+
Scrophulariales	Bignoniaceae	+						+
	Myoporaceae	+						
	Buddlejaceae	+						
	Stilbaceae			+				
	Retziaceae			+				+
	Selaginaceae		+					+
	Globulariaceae	+						
	Plantaginaceae	+						
	Pedaliaceae	+						
	Martyniaceae	+						
	Scrophulariaceae	+	+		+			+
	Lentibulariaceae	+						
	Acanthaceae							+

ture (Corner 1977), and other new data have been considered in the new alignments, although they do not seriously alter the diagram. The affinity between Myrtales and Theales, previously advocated, for example, by Hickey and Wolfe (1975) is not supported by the studies of Corner (1977) or Briggs and Walter (1979), whereas the similarities between Theales and Dilleniales have been stressed more in recent years.

A framework as presented here must always be conceived as a dynamic working model. Individual positions of orders or superorders must not be conceived as ultimate, fixed standpoints. They are relative to those of other groups and may change with new evidence.

Iridoids

A general treatment of the occurrence of iridoids in the angiosperms was given by Jensen, et al. (1975), and their occurrence in the "Tubiflorae" sensu Wettstein has been elucidated recently by Hegnauer and Kooiman (1978).

The occurrence of iridoids in the angiosperms is shown in Figure 2. The following notes can be made on their distribution in the orders of angiosperms, as recognized in the classification presented here.

Iridoids are restricted to the orders Fouquieriales, Sarraceniales, Ericales, Eucommiales, Cornales, Loasales, Dipsacales, Goodeniales, Oleales, Gentianales, Hippuridales, Lamiales and Schrophulariales, besides a few scattered occurrences in the rest of the system, viz. in *Liquidambar* (Altingiaceae or Hamamelidaceae), *Daphniphyllum* (Daphniphyllaceae, of somewhat uncertain affinity, maybe closest to Buxaceae), and *Xylocarpus* (Meliaceae, Rutales). These orders for the most part have sympetalous flowers. The types and distribution of iridoids is presented in Tables 1 & 2 and Figure 3.

Secoiridoids (any of Groups VI-VIII, see Table 2) (Fig. 5) are known to occur in Sarraceniales, Cornales, Loasales, Dipsacales, Goodeniales, Oleales and Gentianales, but not in any other order. In the latter they occur most frequently in the orders of the Gentianiflorae (Goodeniales, Oleales, Gentianales) and Dipsacales of Corniflorae. Oleuropein type compounds are restricted entirely to the Oleales (= Oleaceae). The gentiopicroside group is restricted to the Gentianiflorae, Dipsacales, and Cornaceae, whereas the simple secoiridoids are scattered throughout the above-mentioned orders.

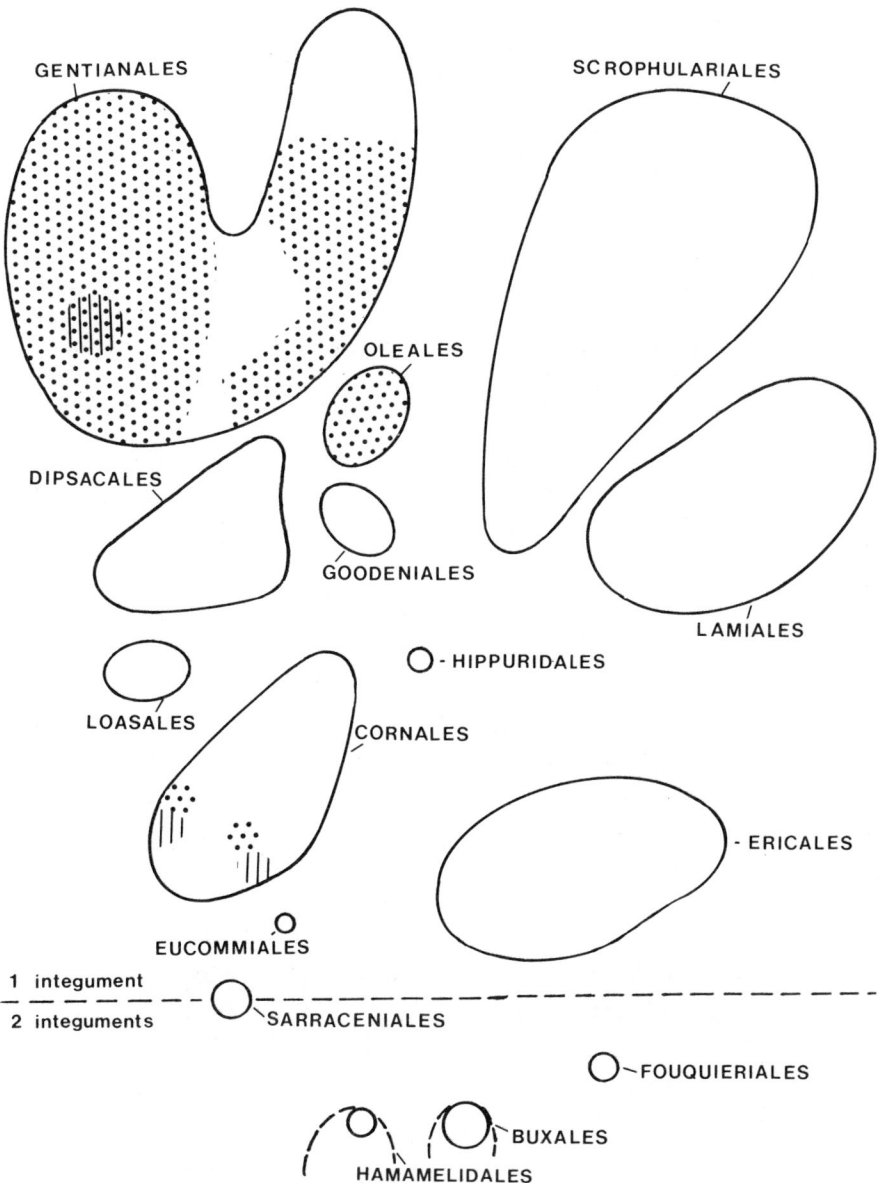

Figure 6. Families containing ipecac alkaloids (vertical hatching), camptothecine, and indole alkaloids of corynanthe, aspidosperma or iboga types (dots) (cf. Fig. 3).

Complex iridoid alkaloids (Fig. 6) can be subdivided into DOPA-derived ipecac alkaloids known in the Alangiaceae, Icacinaceae and Rubiaceae; and indole alkaloids which are found in the Gentianiflorae, viz. Nyssaceae, Icacinaceae, Oleaceae, Loganiaceae, Rubiaceae and Apocynaceae. The few occurrences outside this group (in Sapotaceae, Annonaceae and Ericaceae) must be treated with reserve.

The distribution of iridoids cannot be evaluated fully if it is not compared with the distribution of sympetaly (Fig. 16), unitegmic ovules (Fig. 18), tenuinucellate ovules (Fig. 17), and cellular endosperm formation (except in Gentianales; Fig. 19) (see Dahlgren 1977). In addition, the distribution of terminal endosperm haustoria will prove to coincide to some degree with iridoid distribution. As will be discussed under these attributes later in this chapter, the occurrence of iridoids is correlated in each case with three or four of these characters. An independent origin of several (or even numerous) groups of plants with this combination of attributes is highly unlikely. However, it does not follow that iridoids are necessarily monophyletic. A few scattered genera where this combination of attributes does not occur but where iridoids have been found are *Liquidambar, Daphniphyllum* and *Xylocarpus* (also, in part, *Fouquieria*). These latter genera are probably distantly related to the other groups in which iridoids are frequently found and which seem to form a more or less coherent complex of orders.

Within this main complex there are several families in which iridoids have not yet been detected or have been found only rarely or in certain groups of genera. Among them are the large families Gesneriaceae and Asclepiadaceae, and for example large parts of the Acanthaceae, Rubiaceae and Lamiaceae (for Lamiaceae, see Dahlgren 1979). There are many other families where iridoids have been detected but are of limited occurrrence. Thus, the synthesis of iridoids apparently can be lost easily, which should warn us against evaluating their absence in a family, especially when this family is poorly investigated in this respect. Also, the occurrence and diversity of iridoids in different genera of a family, or even in different species of the same genus, may be very different.

The absence of iridoids in the Asclepiadaceae and their presence in part of the obviously closely related Apocynaceae has been used as a warning against using iridoids for generalizing phylogenetic conclusions. In the light of this, their

Figure 7. Occurrences of anthraquinones. Dots represent acetate derived, and asterisks shikimic acid derived anthraquinones. The diagram should be compared to that of Fig. 2.

absence in the Solanales and Boraginales must not be overemphasized either. These two orders have many of the embryological and morphological attributes characteristic of iridoid-producing orders, and it is not unlikely that the ancestors of Solanales or Boraginales might have had the ability to synthesize iridoids, or that iridoids even may be detected in extant taxa of these orders.

However, the embryological and morphological characters of families such as Actinidiaceae, Sarraceniaceae and Loasaceae and of genera such as *Theligonum* or *Hippuris* indicate that each of these groups have their closest relatives among different taxa with the same properties in the Corniflorae-Gentianiflorae-Lamiiflorae complex. The presence and kinds of iridoids present in each of the above taxa not only support this contention, but contribute additional material for the closest affinities in this complex. In addition, the Fouquieriaceae in the light of its entire combination of attributes, including its possession of iridoids, ellagitannins, polyandrous flowers, tricarpellate gynoecium, etc., in our opinion is more in accord with an Ericalean-Cornalean affinity than with any other (Dahlgren, et al. 1976).

Anthraquinones

In this connection, we will consider the distribution (Fig. 7) and pathways of the anthraquinones in relation to the distribution of iridoids. Anthraquinones which are acetate-derived have a wide and scattered occurrence in angiosperms and are known in the following families: Monocotyledons--Asphodelaceae (*Aloë*), Iridaceae, Zingiberaceae; Dicotyledons--Anacardiaceae, Asteraceae, Chenopodiaceae, Clusiaceae, Combretaceae, Euphorbiaceae, Fabaceae, Lythraceae, Sonneratiaceae, Polygonaceae, Rhamnaceae, Scrophulariaceae (?), Simaroubaceae, Solanaceae, Urticaceae.

Anthraquinones which are shikimic acid-derived are know to occur in certain dicotyledon families only, viz.: Bignoniaceae, Gesneriaceae, Rubiaceae, Scrophulariaceae, Verbenaceae. This distribution does not seem to be random. Except for a dubious record for acetate-derived anthraquinones in Scrophulariaceae, these are not yet recorded in iridoid-containing orders, whereas shikimic acid-derived anthraquinones are known in Gentianales, Lamiales and Scrophulariales, and nowhere else.

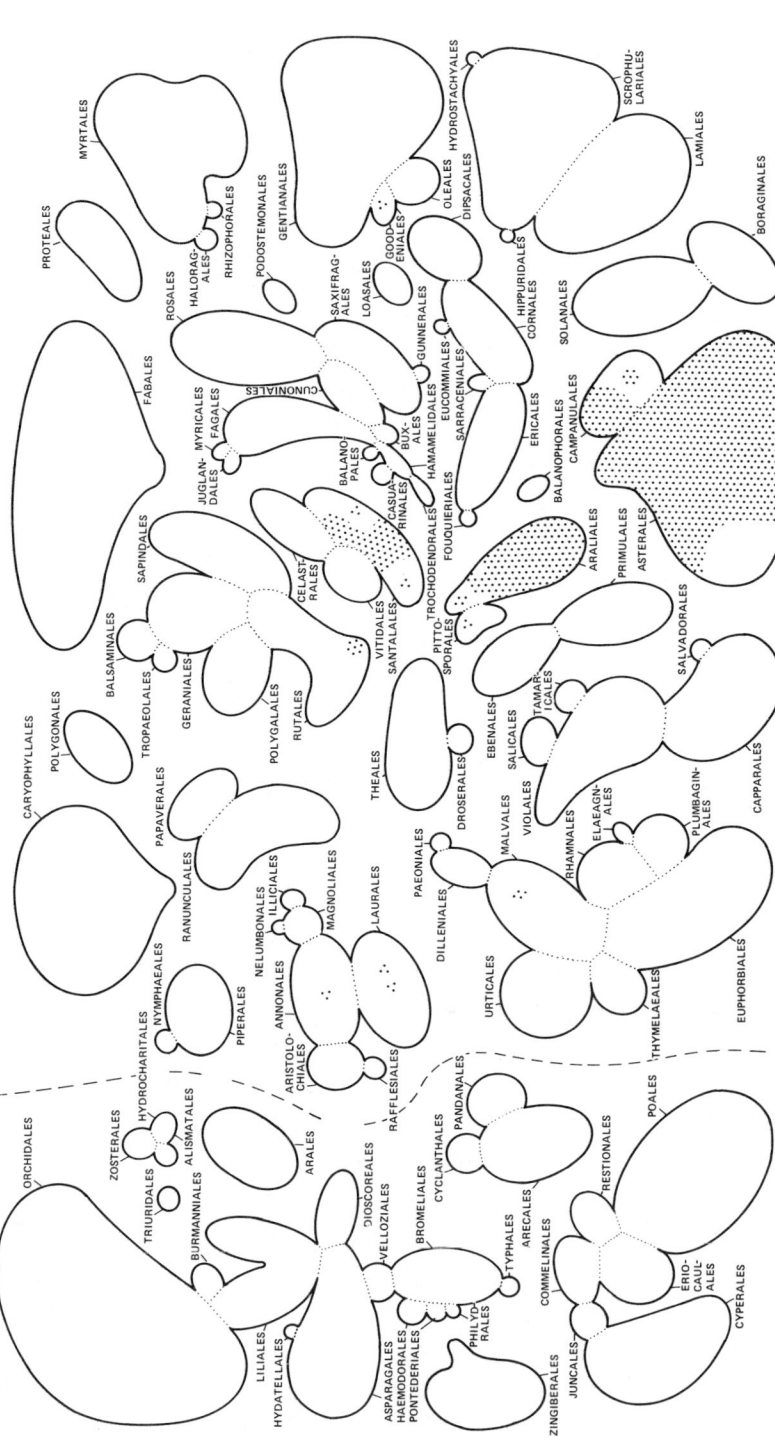

Figure 8. Approximate distribution of polyacetylenes derived from fatty acids. In the "Sympetalae", the polyacetylenes have a distribution largely vicarious to that of the iridoids.

Table 4. Distribution of polyacetylenes (derived from fatty acids) in flowering plants.

Asterales
  Asteraceae (numerous genera: see text)

Araliales
  Apiaceae (numerous genera)
  Araliaceae (*Aralia, Hedera, Panax, Polyscias*)

Pittosporales
  Pittosporaceae (*Pittosporum*)

Campanulales
  Campanulaceae (*Adenophora, Campanula, Codonopsis, Jasione, Plastycodon, Symphyandra, Trachelium, Wahlenbergia*)
  Lobeliaceae (*Lobelia*)

Goodeniales
  Goodeniaceae (*Scaevola*)

Santalales
  Olacaceae (*Olax, Ongokea, Ximenia*)
  Opiliaceae (*Canjera, Opilia*)
  Santalaceae (*Acanthosyris, Antholobus, Choretrum, Comandra, Dendrotrophe, Eucarya, Exocarpus, Leptomeria, Omphacomeria, Pyrularia, Santalum, Thesium*)
  Viscaceae (*Viscum*)
  Loranthaceae (*Nuytsia*)

Rutales
  Simaroubaceae (*Picramnia*)

Malvales
  Sterculiaceae (*Sterculia*)

Fabales
  Fabaceae (*Vicia*)

Laurales
  Lauraceae (*Litsea, Persea*)

Annonales
  Annonaceae (*Alphonsea*)

Dubious:

Euphorbiales
  Euphobiraceae (*Ricinus*)

Dipsacales
  Valerianaceae (*Valeriana*)

Although these data are too few for ultimate conclusions, it seems that the shikimic acid pathway of anthraquinone biosynthesis has evolved within a part of the main evolutionary complex of angiosperms where iridoids are fairly common.

Polyacetylenes (derived from fatty acids)

Polyacetylenic compounds have been studied by various workers, and are summarized by Bohlmann, et al. (1973). Their distribution (Fig. 8) is of interest because to a great extent it is vicarious to the occurrence of iridoids. Polyacetylenes have their richest occurrence in Asteraceae (Compositae) and Apiaceae (Umbelliferae), and their total distribution is presented in Table 4.

In Asteraceae, polyacetylenes show an interesting pattern (Sørensen 1977) in being largely or totally absent in the trible Senecioneae, depending on the circumscription of this tribe. They are by far most varied in the tribles Heliantheae and Anthenideae.

The common occurrence of polyacetylenes in Araliales (Apiaceae and Araliaceae) and in Asterales is now usually considered to express phylogenetic relationships between these orders (Hegnauer 1964, 1977). Floral parts are largely isomerous in the two orders and the dense conglomeration of flowers to an umbel and capitulum, respectively, is similar, with a resultant tendency for zygomorphy of the peripheral flowers (found also in other groups). Stipules are lacking in both groups; the nodes are multilacunar; the pollen grains trinucleate; the ovules unitegmic with at least partly nuclear endosperm formation; endosperm haustoria are lacking, etc. There are also other chemical similarities, such as the occurrence of sesquiterpene lactones and biogenetically similar or identical coumarines, between the two groups. However, there are also great differences, which highly modify this picture. The similarity between Asterales and Campanulales is also supported by the common presence of polyacetylenes. Further, it seems, Pittosporaceae is likely to be linked to Araliales (Hegnauer 1969, p. 330). The occurrence of polyacetylenes also supports the presumed relationships between the Olacaceae, Opiliaceae, Santalaceae, Loranthaceae and Viscaceae, supporting their treatment in one order, the Santalales. The link between this and Araliales is uncertain but likely.

Other records for polyacetylenes mentioned above are difficult to evaluate; probably they have evolved along several, but not numerous, lines of evolution.

Sesquiterpene Lactones

Sesquiterpene lactones are C_{15} terpenoids derived via the mevalonate pathway. They commonly are referred to as bitter principles, which suggests that they may protect the plant by serving as antifeedants at least to mammals. Although it has not yet been proven, they are all probably biogenetically related.

Sesquiterpene lactones are known to occur in the following families of angiosperms, the number of taxa in parentheses (Herz 1977): Asteraceae (c 450); Apiaceae (12); Burseraceae (1); Lauraceae (1); and Magnoliaceae (5). Other lactones (loc. cit.) are known from Amaranthaceae, Aristolochiaceae, Canellaceae and Lauraceae, a surprising concentration in the Magnoliiflorae.

The distribution of sesquiterpene lactones often is referred to as an indication of affinity between Asterales and Araliales (in which Apiaceae is included), and (although with much less certainty) as a further indication of affinity with the Rutiflorae: Rutales. Further investigations in Araliaceae, Campanulaceae and other families should be of great interest. The occurrence of sesquiterpene lactones in the tribes of Asteraceae is of taxonomic interest. They are common and complex in Heliantheae (incl. some Helenieae); of intermediate frequency and complexity in the Anthemideae, Cichorieae, Cynareae, Senecioneae, Inuleae and Vernonieae; infrequent and biogenetically less advanced in the Eupatorieae; and finally not known in the Mutisieae, Astereae, Tageteae and Arctoteae-Calenduleae.

It should be mentioned that sesquiterpene lactones have been reported in the literature to occur in Scrophulariaceae, but these few reports are due to taxonomic mistakes (e.g., *Veronica* for *Veronia*).

Glucosinolates

Glucosinolates comprise a natural group of sulphur-con-

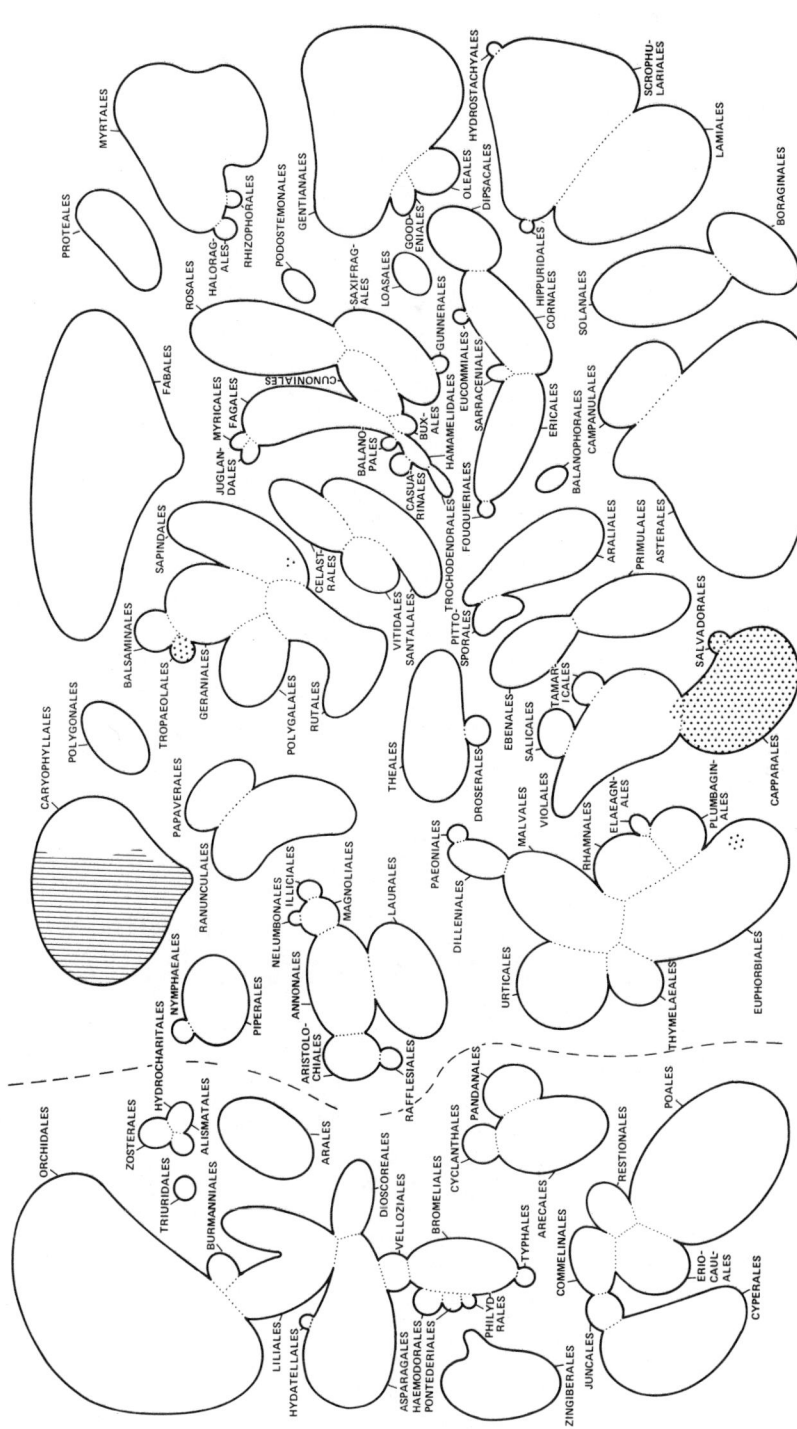

Figure 9. Approximate distribution of glucosinolates (dots) and of betalains (hatching).

taining glucosides, which by a complex of enzymes, called myrosinase, form mustard oils or isothiocyanates. Their occurrence is still under investigation (Ettlinger, Kjaer, Rodman, etc.), but is concentrated in particular families, in most of which glucosinolates are invariably present (Fig. 9). The glucosinolates comprise a variable group of compounds, the biosynthesis of which is likely to have evolved independently in a few lines of evolution. They occur in the following groups of the present classification:

> Capparales: Capparaceae (incl. Cleomaceae, Pentadiplandraceae and Koeberliniaceae), Brassicaceae (=Cruciferae), Tovariaceae, Resedaceae, Moringaceae, Gyrostemonaceae, Batidaceae
> Salvadorales: Salvadoraceae
> Violales: Caricaceae
> Euphorbiales: Euphoribaceae: *Drypetes* ("Putranjivaceae")
> Tropaeolales: Tropaeolaceae, Limnanthaceae

Besides, glucosinolates probably are present in Bretschneideraceae, which is often placed in Sapindales, a position which is here accepted with some reluctance. Another probable case is *Akania* (Akaniaceae, Violales), in which the presence of glucosinolates needs to be verified.

According to Ettlinger (pers. com.), the glucosinolates of four of the families of the Capparales (Tovariaceae, Resedaceae, Brassicaceae, Capparaceae) can be classified as "advanced" (with tryptophane-derived glucosinolates) with successively less specialization occurring along the sequence of families given above. Moringaceae is dubiously allied to the other families in Capparales. It seems to contain "primitive" glucosinolates. Kolbe (1978) on the basis of serological results also preferred not to include Moringaceae in the Capparales, but to treat it separately as the order Moringales.

Batidaceae and Gyrostemonaceae may or may not deserve a place in the Capparales. Thorne (1976) allies Gyrostemonaceae with Sapindaceae (Sapindales) and places Batidaceae in his "incertae sedis" list. Goldblatt, et al. (1976) prefer to place the two families together in a separate order in the vicinity of Capparales, which has also been accepted by Cronquist (1979). However, some floral and other morphological similarities between Gyrostemonaceae and Resedaceae (supported by anatomical similarities (L. Bolt - Jørgensen, pers. com.)) indicate that the position of the families in Capparales (as in Dahlgren 1975a) may be justified.

Salvadoraceae has simply constructed, haplostemonous, often tetramerous flowers with nectar glands. The carpels are two in number and form a drupe or berry. There is no clear indication of the relationships of this family, which frequently is placed with Celastraceae in Celastrales, a dubious position. We have preferred to place it separately near Capparales until further evidence provides a basis for a more definitive affinity.

While Caricaceae is normally placed in Violales (or the corresponding order), *Drypetes* (incl. *Putranjiva*) still should be reconsidered as to whether or not is belongs in the somewhat heterogenous Euphorbiaceae.

Limnanthaceae and Tropaeolaceae show mutual similarity in several features, but also differ in some details. Chemically, Limnanthaceae is interesting in that it forms ellagic acid, in which it differs from all (or at least the vast majority) of the other glucosinolate-producing plants. The position of these two families in or near Capparales is dubious, although not wholly unlikely. Boulter, in his phylogenetic tree (Boulter 1973a, p. 548, Fig. 5), had *Tropaeolum* in close vicinity to *Brassica*, although on the basis of the same results the genera appears, for unknown reasons, at different ends in another phylogenetic tree published the same year (Boulter 1973b, p. 213). The genera of Tropaeolales show floral similarities with Geraniales and are placed in conjunction with them until further data can support a different position. Serological results (Kolbe 1978) did not show affinity between Tropaeolaceae and some members of Capparales investigated.

Betalains

Betalains have been surveyed by Mabry in various publications (e.g., 1966, 1976). In the angiosperms, they are wholly restricted to the Caryophyllales. They occur in most of the families of this order, but not in Caryophyllaceae and Molluginaceae, where they are replaced by anthocyanins. In spite of this, Caryophyllales (after the exclusion of Theligonaceae,

Batidaceae, Gyrostemonaceae, Polygonaceae and various other families) form a natural group characterized by several embryological features (e.g., the mostly campylotropous, bitegmic ovules with nuclear endosperm formation; frequently caryophyllad type of embryogenesis; curved embryo and a richly developed perisperm; by the frequently free central placentation); by anatomical features (e.g., the frequent occurrence of aberrant secondary thickening growth); by pollen morphological properties; and above all by a very peculiar type of sieve tube plastid.

The occurrence of anthocyanins in the group is partly correlated with the occurrence of a polygonal crystalloid in the center of the plastid in Caryophyllaceae, a type which also occurs in Molluginaceae, but also in *Stegnosperma* (Behnke 1976a, b; Mabry 1977). The last mentioned genus is a betalain plant. It deviates from all the members of Phytolaccaceae, where it is often placed, in having petals as well as sepals, as in the Caryophyllaceae. It also has a similar capsular fruit. Whether the similarities between *Stegnosperma* and Caryophyllaceae indicate phylogenetic affinity or have evolved by convergence is yet uncertain, but it is clear that the embryological evidence and other characters in combination with the very typical annular structure of protein filaments in the sieve tubes indicate that Caryophyllales are a coherent group, and that the betalain/anthocyanin border cuts this in a particular way (see Behnke 1976b and Mabry 1976). It often has been indicated that Phytolaccaceae s. lat. is primitive by virtue of its nearly apocarpous gynoecium (*Phytolacca*) and its suggestive variation trends through *Stegnosperma* in the direction of Caryophyllaceae; through *Seguieria* to Pereskioideae of Cactaceae; through *Rivina* to the monocarpellate Nyctaginaceae; through *Microtea* to Chenopodiaceae. But here again we meet with problems of convergence versus a true affinity. One may also argue that petalless types of anthocyanin Caryophyllaceae could be basic, having evolved into a betalain line of evolution which soon differentiated in the various main directions. However, from a taxonomic point of view, the evolutionary sequence is rather immaterial. A third and very likely possibility is that the betalain and anthocyanin families evolved along parallel lines from ancestors where floral pigments were not yet widespread (Mabry 1976).

Ellagic Acid and Ellagitannins

The distribution of the polyphenolic compound ellagic

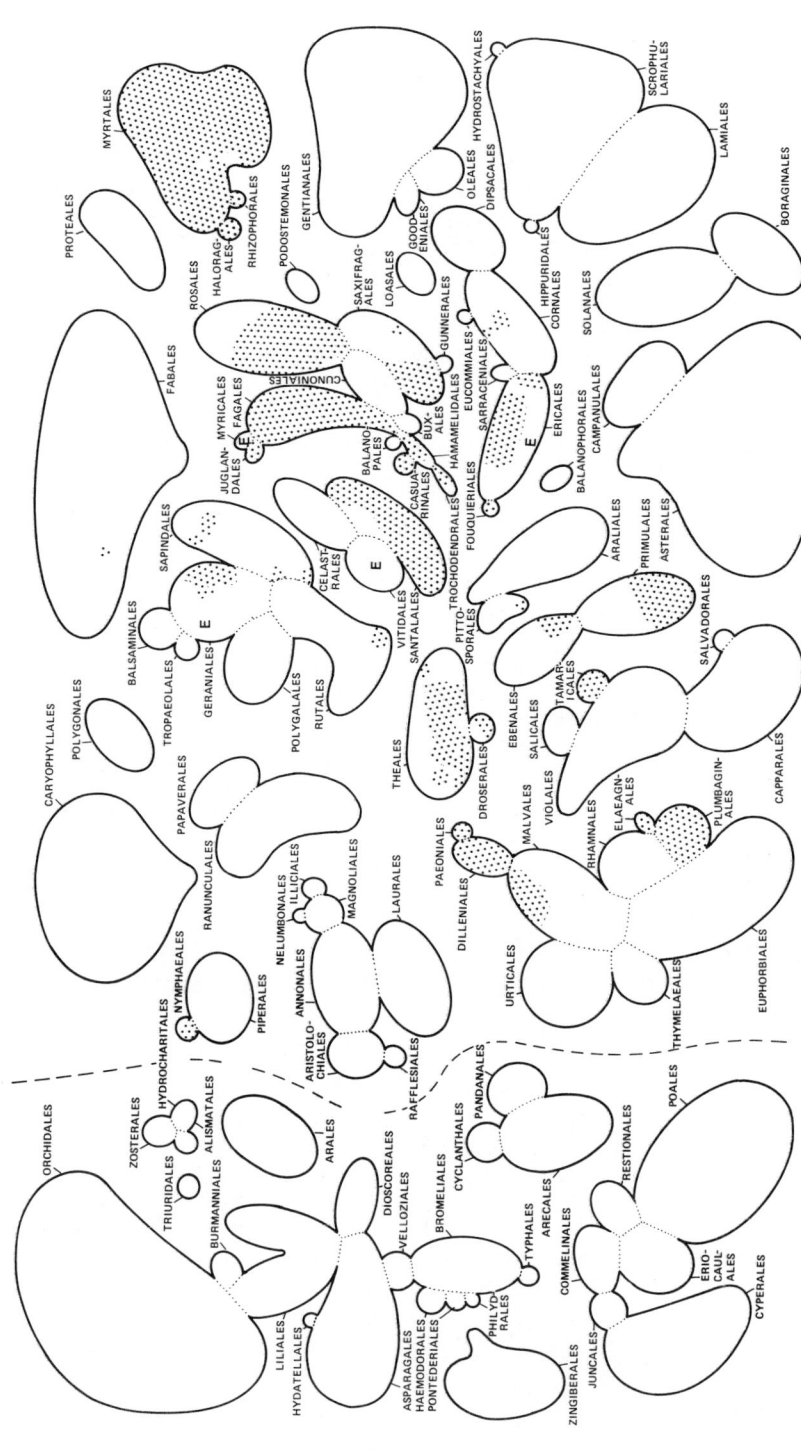

Figure 10. Approximate distribution of ellagi-tannins (dots) and ellagic acid (when ellagic acid only is reported, the dot shading is substituted with an E). Data chiefly from Bate-Smith (pers. comm.).

acid has been described and shown in various diagrams before (Kubitzki 1969; Bate-Smith 1962, 1972, 1974) and are shown in Dahlgren (1977a) (in the classificatory diagram of Dahlgren, 1975). The data plotted in this diagram as well as in that presented (Fig. 10) here have kindly been supplied by Dr. Bate-Smith.

Ellagitannins show the most conspicuous features by the main groups where they are absent, viz.

(1) in all monocotyledons;
(2) in the Magnoliiflorae and Ranunculiflorae (which often have benzylisoquinoline alkaloids);
(3) in the Caryophylliflorae;
(4) in the Araliiflorae-Asteriflorae complexes;
(5) in the Solaniflorae;
(6) in Gentianiflorae-Loasiflorae-Lamiiflorae complexes (representing most of the iridoid plants);
(7) in most Fabiflorae and all Proteiflorae;
(8) in most orders of the Violiflorae and Rutiflorae, although with certain exceptions.

Centers of ellagitannins are the Rosiflorae complex, which here includes Hamaelidales, Fagales, Juglandales, Myricales, Casuarinales, Cunoniales, Saxifragales, and Rosales. In addition, ellagitannins are common in the Myrtiflorae and the Theiflorae, and scattered in the Dilleniflorae, which includes, for example, the Dilleniales, Malvales, Elaeagnales and Plumbaginales.

Ellagitannins probably evolved in several lines of evolution. Where they occur they are often inconsistently present, and more common in woody than in herbaceous plants. In Rosales, they occur within Rosaceae s. str., but are obviously absent in Malaceae and Amygdalaceae (which synthesize cyanogenic compounds)

One group where the presence of ellagitannins is somewhat unexpected in dicotyledons is the Nymphaeales (Nymphaeiflorae). This group shows many similarities to the monocotyledonous Alismatiflorae, but is no doubt more closely allied to the Piperales. The Piperales, again, show affinity to the Annonales and Magnoliales, having cells with essential oils as do these orders. Piperales mostly lacks benzylisoquinoline alkaloids (see below) and has a copious starch perisperm just as have the Nymphaeales. It is likely that Piperales and Nymphaeales developed within the Magnoliiflorean ancestral

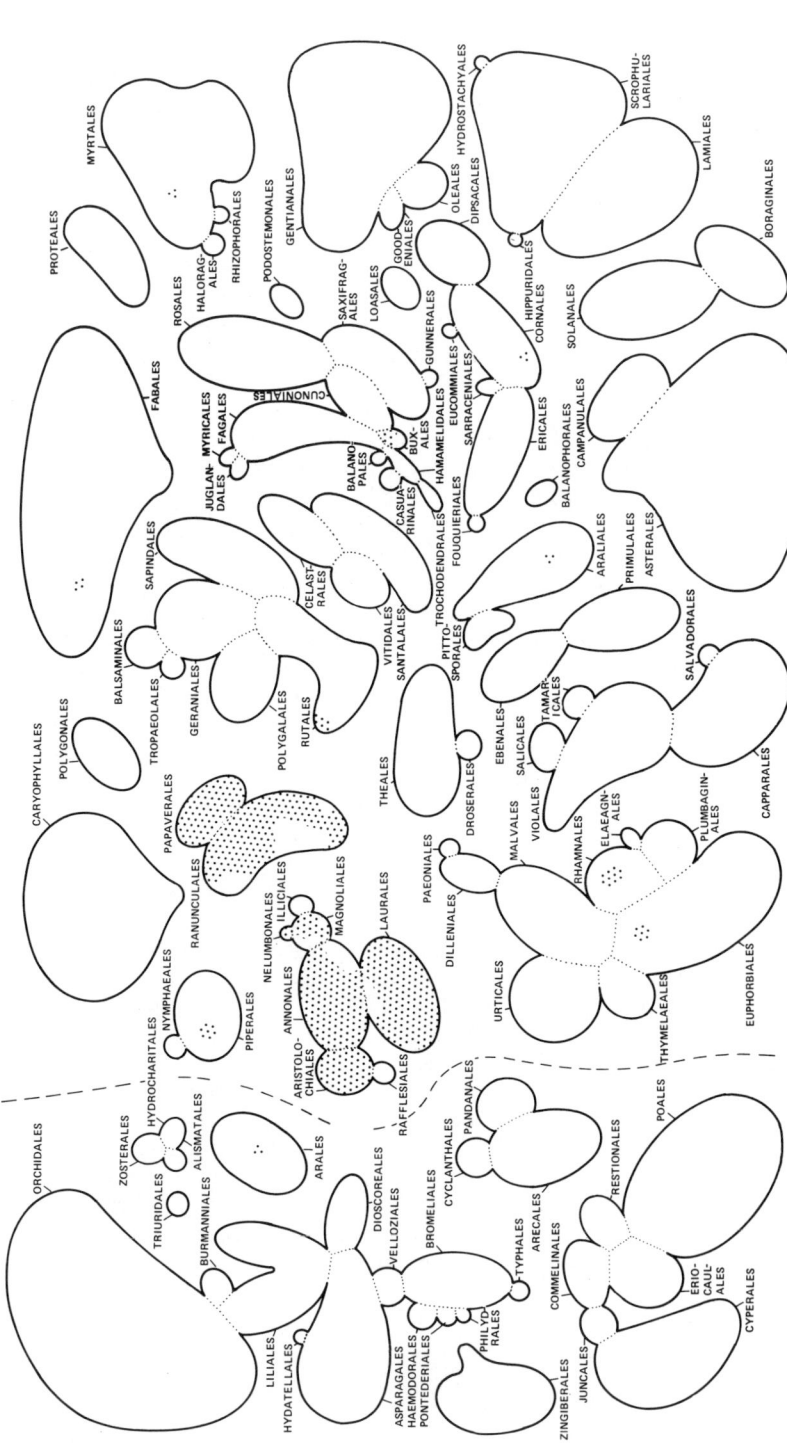

Figure 11. Approximate distribution of benzylisoquinoline alkaloids (and closely related alkaloids).

stock and, in being adapted to forest and aquatic life respectively, they evolved the different, specialized floral structures and also (especially Nymphaeales) went through a chemical evolution divergent from that of the Magnoliiflorae.

Benzylisoquinoline Alkaloids

A considerable literature has accumulated about the benzylisoquine alkaloids and their distribution (Fig. 11), and only a brief comment needs to be made here. In the past few decades, it became clear that the former order "Rhoedales" consisted of two distantly related groups of families, one of which is characterized by benzylisoquinoline alkaloids (Papaverales in the present sense), the other by glucosinolates (Capparales in the present sense). In spite of being acknowledged as separate groups, they exhibit surprising similarities in floral morphology (cf. for example *Hypecoum* and other Fumariaceae with *Cleome* and other Capparaceae), a resemblance due to convergence. While flowers of the capparalean families seem to have evolved from a 4- or 5-merous basic pattern, those of the benzylisoquinoline alkaloids containing Papaverales have a basically 3- or 2-merous pattern. Trimerous flowers are found, for example, in *Platystemon* and *Argemone*. The latter closely approaches in symmetry the types found in the families of Ranunculales, for example in Berberidaceae. The affinity between Papaverales and Ranunculales is indeed so great that these orders could with justification be treated as one, as done by Thorne (1976) with this Berberidales.

Within the Magnoliiflorae, beynzylisoquinoline alkaloids are distributed among the orders Magnoliales, Annonales, Aristolochiales, Laurales and Nelumbonales, but are not known to occur in Chloranthales, Illiciales or Rafflesiales. They also occur very rarely in the Piperales, where they are, however, reported only for single species of *Piper*. The benzylisoquinoline alkaloids in the Magnoliiflorae are much less complex than those in Papaverales. Their occurrence and variety in Ranunculaceae have been used, along with features (e.g., serological tests, the type of fruit, the presence of honey tepals, chromosome characters, and the occurrence of other chemical constitutents, such as diterpenes, alkaloids, protoanemonin and cyanogenic compounds) by Jensen (1968) and Frohne and Jensen (1979) as the basis of subdividing the family.

Figure 12. Approximate distribution of tropane alkaloids (hatching) and pyrrolizidine alkaloids (dots), according to Romeike (1978) and Culvenor (1978), respectively.

Table 5. Distribution of tropane and pyrrolizidine alkaloids.

Tropane alkaloids:
- Convolvulaceae: *Calystegia, Convolvulus*
- Cruciferae: *Cochlearia*
- Dioscoreaceae: *Dioscorea*
- Elaeocarpaceae: *Peripentadenia*
- Erythroxylaceae: *Erythroxylum*
- Euphorbiaceae: *Phyllanthus*
- Orchidaceae: *Dendrobium*
- Proteaceae: *Bellendena, Knightia, Darlingia*
- Rhizophoraceae: *Bruguiera, Carallia, Gynotroches*
- Solanaceae: many genera

Pyrrolizidine alkaloids:
- Apocynaceae: *Anodendron, Parsonia, Urechites*
- Asteraceae: *Cacalia, Eupatorium, Senecio*
- Boraginaceae: numerous genera
- Celastraceae: *Bhesa*
- Chenopodiaceae: *Anabasis*
- Elaeocarpaceae: *Aristotelia*
- Euphorbiaceae: *Phyllanthus, Securigera*
- Fabaceae: *Adenocarpus, Crotalaria, Cytisus, Laburnum*
- Orchidaceae: several genera
- Poaceae: *Festuca, Lolium, Thelepogon*
- Ranunculaceae: *Caltha*
- Rhizophoraceae: *Cassipourea*
- Santalaceae: *Thesium* (semiparasitic)
- Sapotaceae: *Mimusops, Planchonella*
- Scrophulariaceae: *Castilleja* (semiparasitic)

The occurrence of benzylisoquinoline alkaloids in the
Zanthoxylum complex (Rutaceae) was discussed by Fish and
Waterman (1973), who regarded the family as derived from
Ranunculiflorean ancestors, whereby this complex would be the
only one in the family (and the superorder Rutiflorae!) that
has retained benzylisoquinoline alkaloids.

The isolated occurrence of benzylisoquinoline alkaloids
and related compounds in *Croton* (Euphorbiaceae), many Rhamnaceae, Symplocaceae, Buxaceae, etc., are even more difficult
to evaluate, as any indications of interrelationship or affinity
to the above-mentioned groups are most dubious. The presence
of benzylisoquinoline alkaloids in *Heracleum* (Apiaceae) (Gupta,
et al. 1976) may be of interest in connection with their occurrence in Rutaceae.

Another important point is the absence of benzylisoquinoline alkaloids in the Nymphaeales (Nymphaeiflorae), a group
which shows much affinity with Piperales, where benzylisoquinoline alkaloids are known to occur but are rare. The
Nymphaeales also have affinities with the Magnoliiflorae.
It is not improbable that these two orders have a common origin, which may have been benzylisoquinoline-containing plants
allied to a Magnoliiflorean ancestral stock.

Tropane and Pyrrolizidine Alkaloids

These groups (Fig. 12) were surveyed by Romeike (1978)
and Culvenor (1978), respectively. The occurrence of these
alkaloids is presented in Table 5. The distribution of each
of these types of alkaloids, judging from their wide and scattered distribution in the angiosperms, cannot be used for any
far-reaching conclusions with regard to affinities between
families or orders. Although the affinity between Solanaceae
and Convolvulaceae is possibly supported by their shared ability
to produce tropane alkaloids, these alkaloids probably can
be used with more success in taxonomy at infrafamiliar levels.

SOME CHEMICAL CHARACTERS IN THE MONOCOTYLEDONS

The distribution of some groups of compounds in monocotyledons given below are taken mainly from Dahlgren and Clifford
(in preparation), and S. R. Jensen and B. J. Nielsen have contributed much of the chemical information.

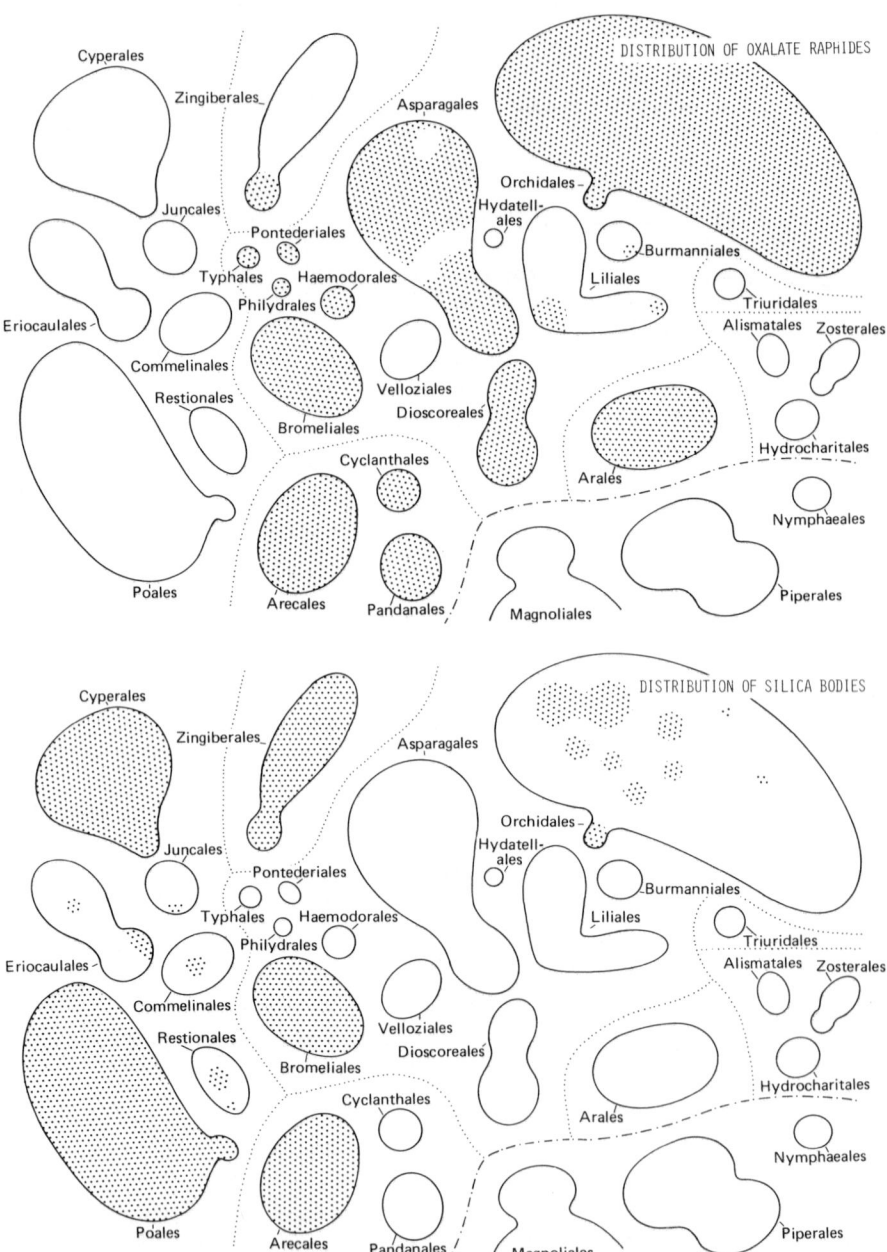

Figure 13. Approximate distribution of oxalate raphides and silica bodies (upper and lower diagrams, respectively) in the monocotyledons.

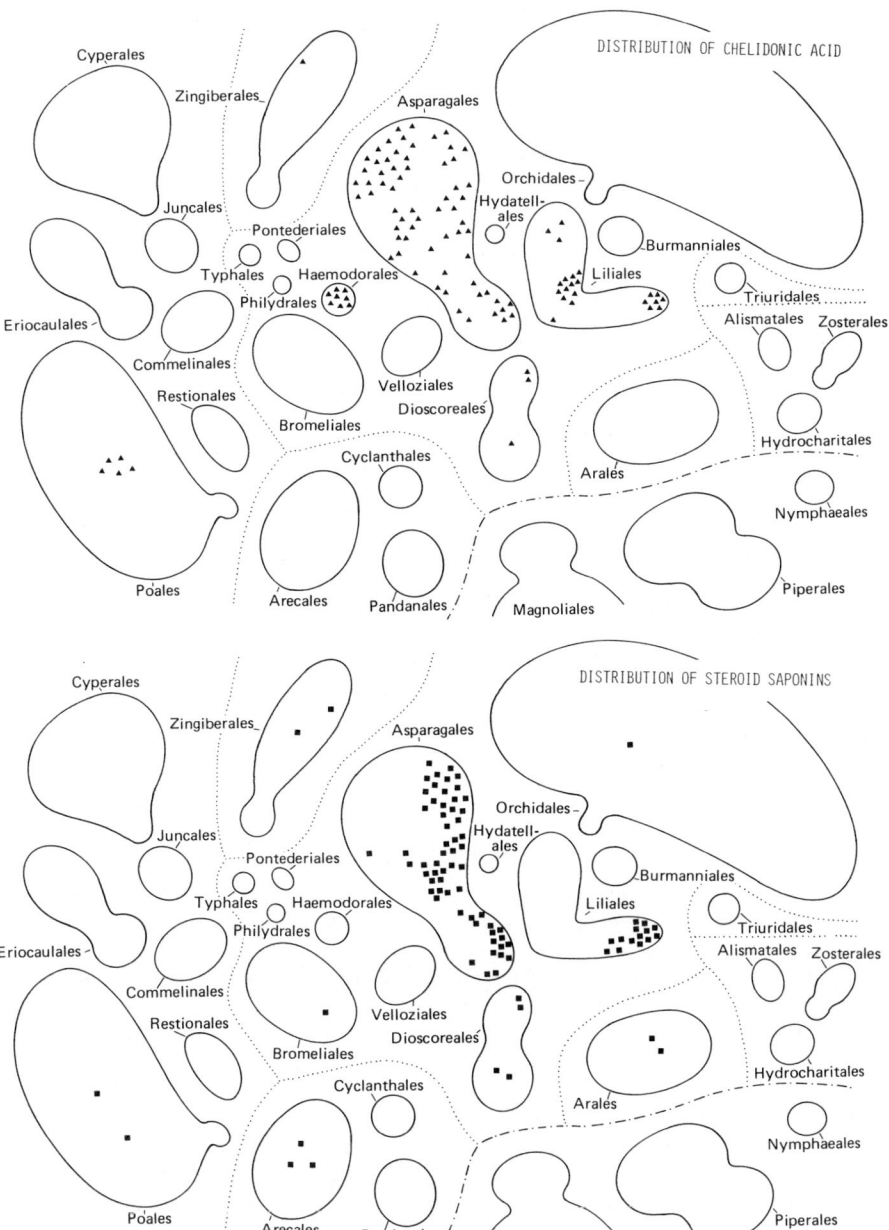

Figure 14. Approximate distribution of chelidonic acid and steroid saponins (upper and lower diagrams, respectively) in the monocotyledons.

Probably well known by now is the distribution of oxalate raphides (Fig. 13, top), i.e., thin crystal needles of calcium oxalate ordered in bundles and deposited in cells, sacs or vessels filled with mucilage. Oxalate raphides are lacking in the Commeliniflorae and Alismatiflorae as defined here, but are of almost universal occurrence in the Ariflorae and Areciflorae. They are very common and have an interesting distribution in the Liliiflorae, although rare in the order Liliales. In the Zingiberiflorae, oxalate raphides occur in families with 5-6 functional stamens, but are lacking in those with one or a half functional stamen.

The distribution of silica bodies (Fig. 13, bottom) is almost equally elucidative. Silica bodies are scattered over most major groups of the Commeliniflorae and Zingiberiflorae, and also are found in the palms (Arecales), Bromeliales, and in many tropical orchids (Orchidales). The distribution and shape of the silica bodies in the Commeliniflorae give some useful indications. Their nearly total absence in Juncales and their very particular shapes in the closely allied Cyperales suggests that they may have originated independently in the latter order and in grasses. It also is likely that they appeared separately in other evolutionary lines, at least in Bromeliales and Orchidales, and perhaps also in other major groups.

Chelidonic acid has the distribution shown in Figure 14 (top). The distribution is concentrated mainly in some orders of the Liliiflorae: Dioscoreales, Asparagales, Liliales, Alstroemeriales and Haemodorales, where chelidonic acid seems to be particularly common in such families as Amaryllidaceae, Hyacinthaceae, Asphodelaceae, Convallariaceae, Colchicaceae, Melanthiaceae and Haemodoraceae, but possibly missing in, for example, Liliaceae s. str. The occurrence of chelidonic acid, although incompletely known, does not seem to be considerable in the other superorders.

Steroid saponins have the very approximate distribution seen in Figure 14 (bottom). This distribution is reminiscent of that for chelidonic acid, but it is noteworthy that steroid saponins are absent or at least very rare in the alkaloid-rich families Amaryllidaceae and Colchicaceae, and in the alkaloid-containing Dioscoreaceae (Seigler 1977). Shown here are only the reports where the steroid saponins have been verified with a certain degree of certainty. Saponins on the whole have a wider distribution, but some of these are no doubt of tri-

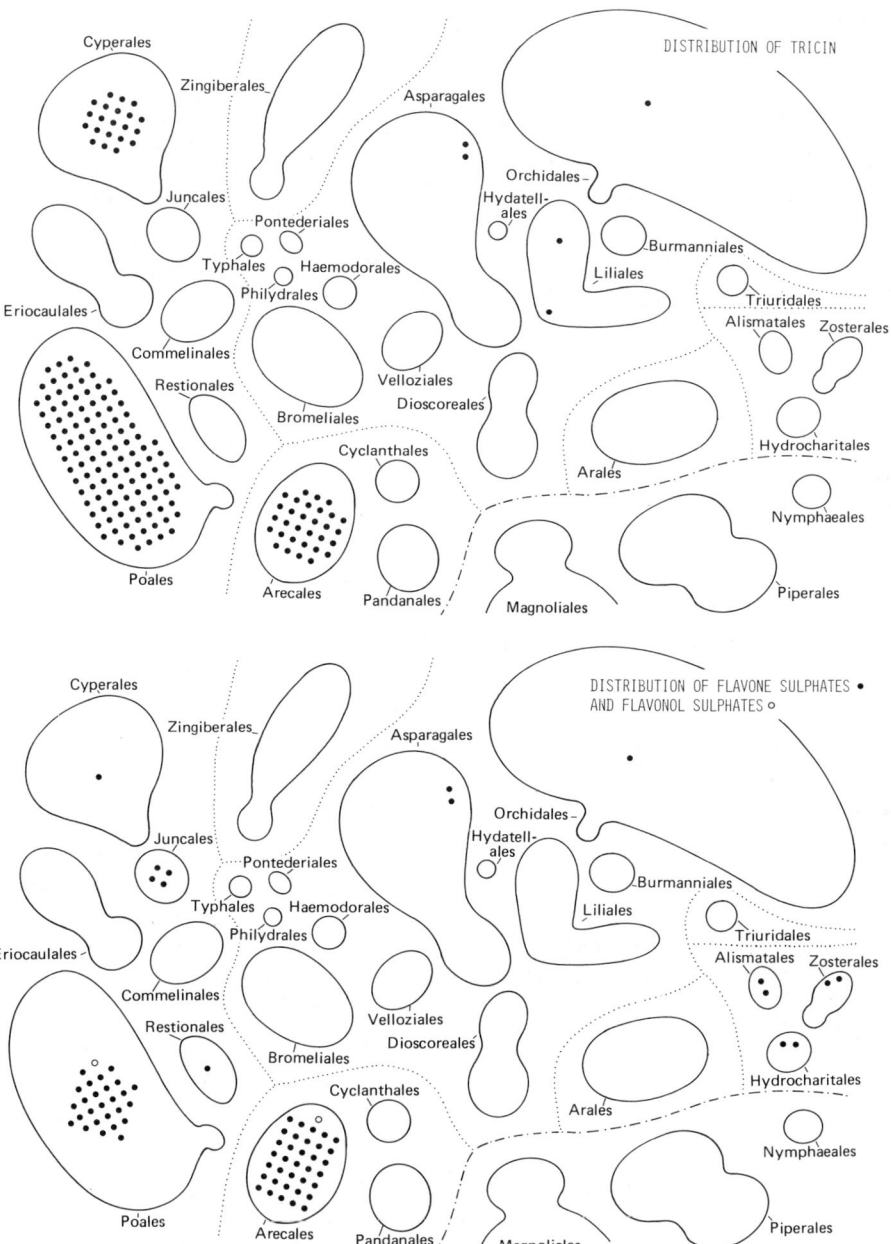

Figure 15. Approximate distribution of tricine and flavonoid sulphates (upper and lower diagrams, respectively) in the monocotyledons (data from Gornall, et al. 1979).

terpene type.

Among the flavonoids (Gornall, et al. 1979), some show a conspicuous distribution in the monocotyledons, for example tricin (Fig. 15, top) and flavonoid sulfates (Fig. 15, bottom). The most conspicuous characteristic for these types of compounds (also supported by a few other groups of flavonoids) is that they occur in Juncales-Cyperales, Poales and Arecales. The puzzling fact that they indicate a possible connection between grasses and palms was pointed out by Harborne (1973).
A possible connection between these was proposed quite independently by Clifford (1967) on the basis of a computer-generated study. However, later studies (Dahlgren and Clifford, in preparation) do not equivocally support such an affinity.

These data indicate that there has been a chemical differentiation in the monocotyledonous ancestors which has kept pace with morphological differentiation. Except for the rather unexpected indication of a Arecalean-Poalean connection, the chemical evidence considered here does not involve any surprising features. A connection with ecological conditions is probable, although not altogether clear from the data presented.

EXAMPLES OF CORRELATION BETWEEN CHEMICAL AND OTHER PROPERTIES

The distribution of each group of chemical compound, like the distribution of other properties, can be used to a variable degree as the basis of a system of classification. However, there are great risks for overemphasizing certain attributes and neglecting others. The following examples are of interest mainly in elucidating the correlation between some chemical properties and other, mostly morphological and embryological attributes. After evaluating the taxonomic value of the distribution of benzylisoquinoline alkaloids, we will proceed to evaluate in more detail that of the iridoid compounds against the distribution of some supposedly independent morphological properties.

When judging the distribution of benzylisoquinoline alkaloids, there has long been agreement about the conclusion that their distribution to a great extent reflects phylogenetic relationships; the families in Magnoliales, Annonales, Aristolochiales, Laurales, and Piperales also are related because of possession in some or all of them of such attributes

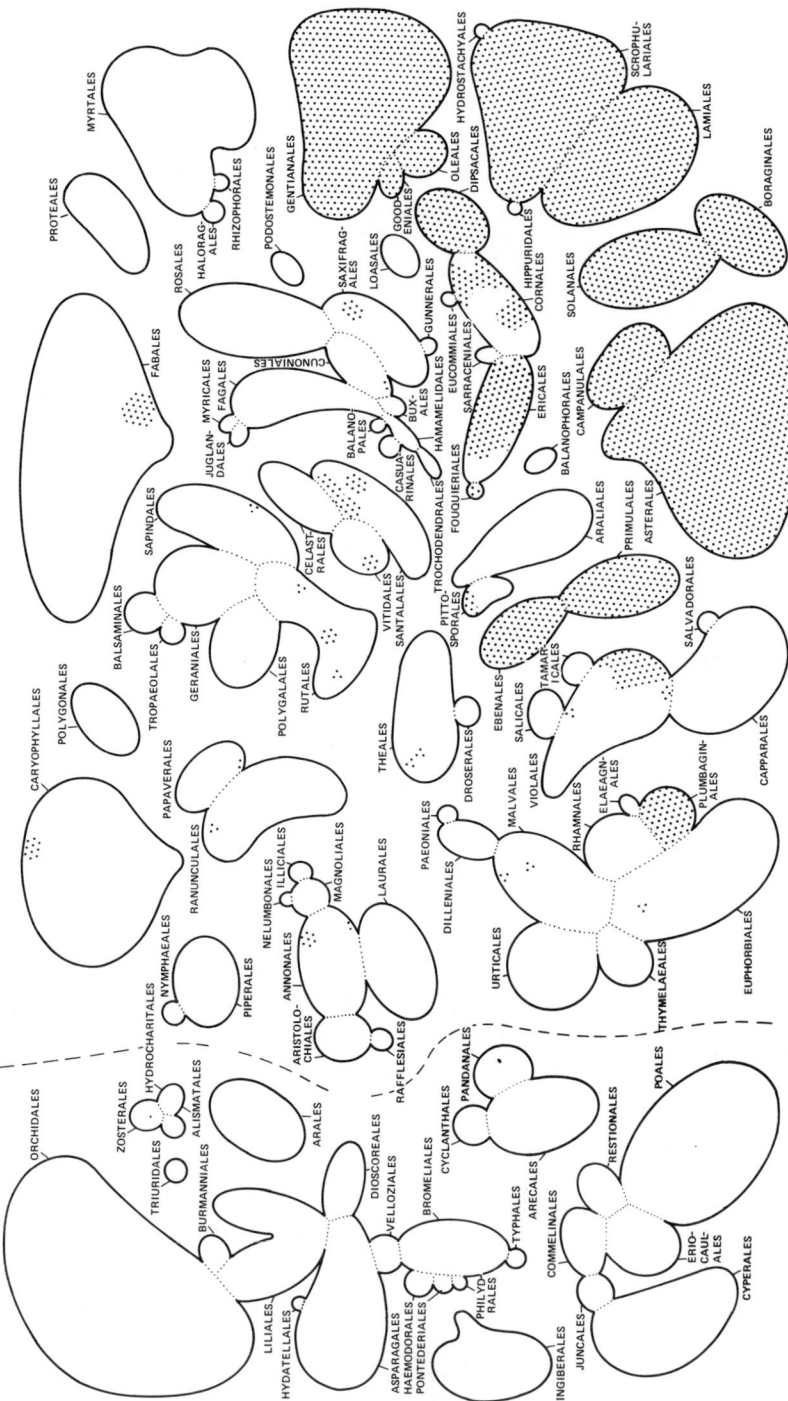

Figure 16. Approximate distribution of sympetalous flowers in angiosperms. Perianth fusion in monocotyledons is not included. Fusion between two petals or pairs of petals as in Fabaceae, Balsaminaceae or Polygalaceae is not included. Homology problems, as to what corresponds to petals, make the shading in, for example, the families of the Santalales somewhat doubtful.

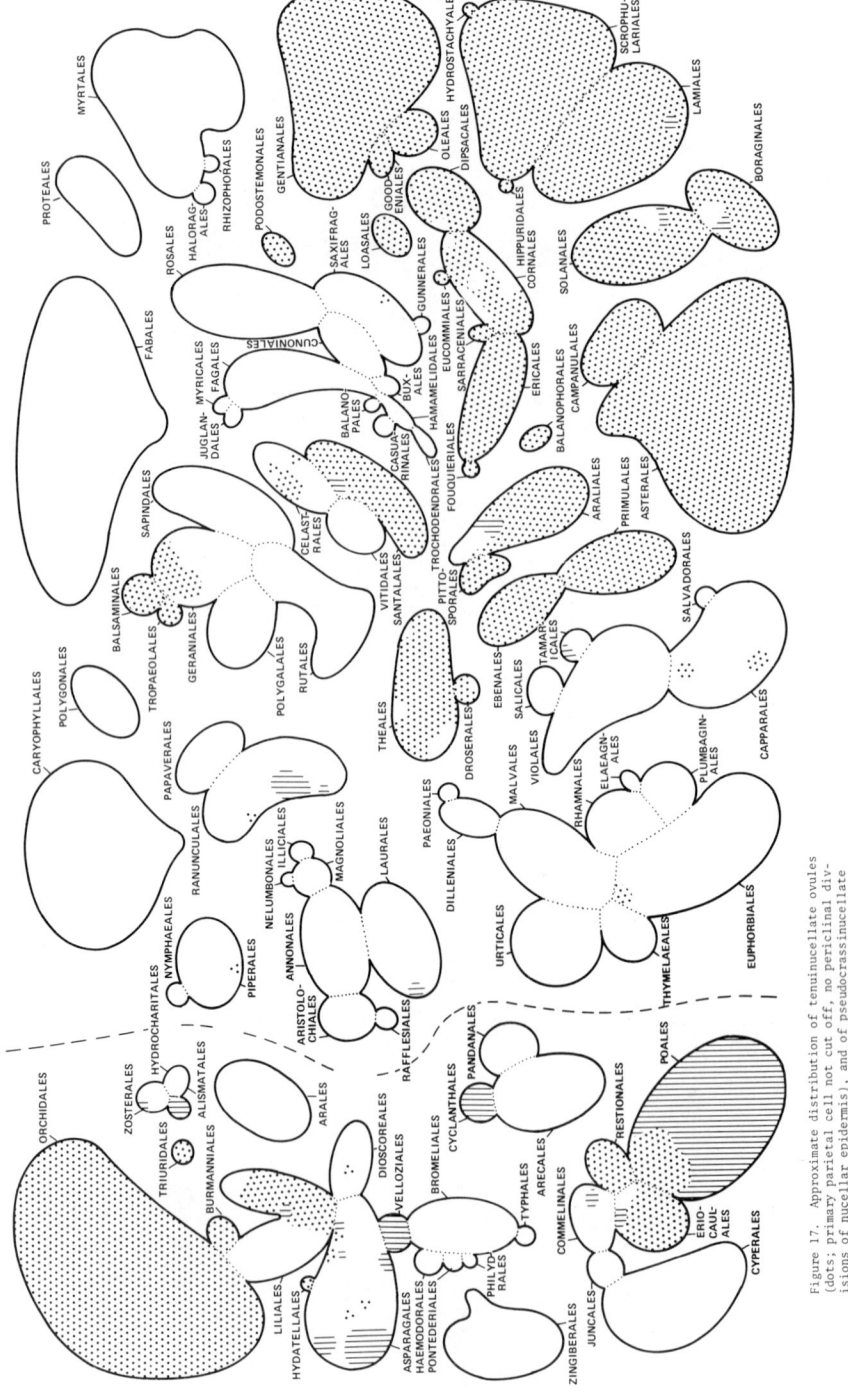

Figure 17. Approximate distribution of tenuinucellate ovules (dots; primary parietal cell not cut off, no periclinal divisions of nucellar epidermis), and of pseudocrassinucellate ovules (hatching; no parietal cell, but divisions in nucellar epidermis to form a "cap" above the embryo sac). Unshaded groups have crassinucellate ovules or (rarely) are unknown.

as:

> monocolpate or inaperturate pollen grains;
> cellular endosperm formation
> cells with essential oils
> apocarpy
> primitive vessel characters
> frequent possession of P-type sieve tube
>   plastids.

Ranunculales and Papaverales have a basic 3-zono-colpate pollen type, and they mostly have nuclear endosperm formation; they also lack cells with essential oils and have simple perforation plates in the end walls of their vessels. Finally, they always possess S-type sieve element plastids. Thus, they immediately stand out as different from the basic Magnoliiflorae, although apocarpy is still a common attribute in both superorders. Certain groups like Illiciaceae, Schisandraceae and Nelumbonaceae bridge this gap, and most taxonomists tend to acknowledge the affinity between the Magnoliiflorae and Ranunculiflorae. More dubious is the connection between these superorders and other benzylisoquinoline alkaloid-containing groups, and subjective judgements must be given.

In the case of the groups with a sympetalous perianth (Fig. 16), these form one fairly concentrated block in the right part of our diagram. This, of course, does not prevent numerous other groups from also being sympetalous, which may be an attribute likely to develop by convergence. We find it, for example, in many Mimosaceae, Crassulaceae, Cucurbitaceae and some Rutaceae, groups which are very obviously unrelated to the conventional "Sympetalae".

It also is dubious that Plumbaginaceae, Primulaceae, Ebenaceae and certain other families are closely related to the "Sympetalae" in the strict sense. How homogeneous the group is could be tested if a fair number of other attributes, judged to be phylogenetically important, are plotted in our diagram.

Tenuinucellate ovules (Fig. 17) have a distribution largely coinciding with sympetaly, admitting the Primuliflorae but not the Plumbaginaceae to belong to this category. However, we have connections outside the classical "Sympetalae" to consider, e.g., in the Araliiflorae and the Corniflorae, which are useful for adjudging the affinities of sympetalous groups.

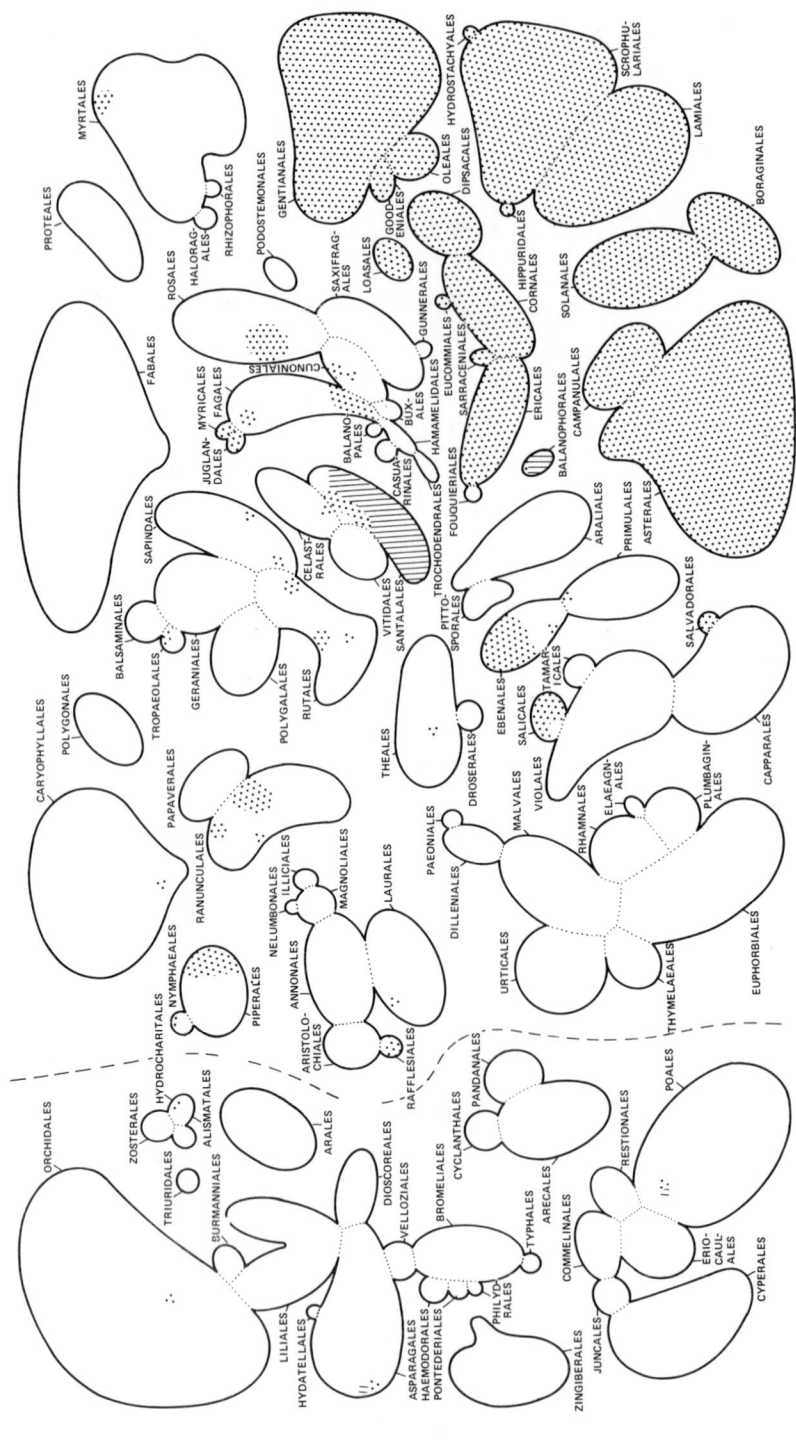

Figure 18. Approximate distribution of unitegmic (dots) and ategmic (hatching) ovules. Unshaded groups have bitegmic ovules.

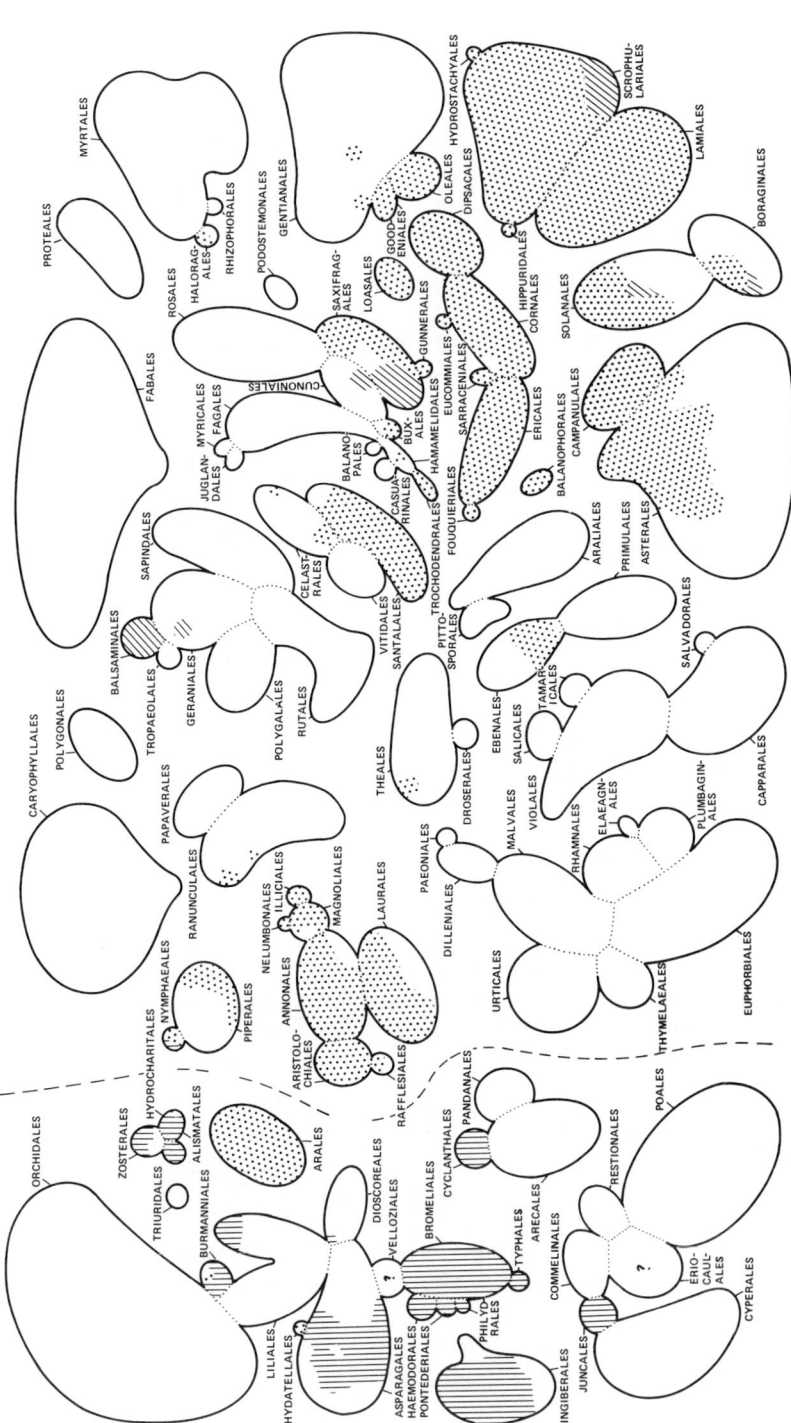

Figure 19. Approximate distribution of cellular (dots), helobial (vertical hatching) and intermediate (oblique hatching) endosperm formation. Unshaded areas mainly denote nuclear endosperm formation, but also include some groups unknown in this respect.

The distribution of ovules with one integument (Fig. 18) gives a similar and even stronger differentiation of the groups considered, and again connects sympetalous groups with the Cornales. The Ericales agree in all three respects (one integument, tenuinucellate condition and sympetaly) with the strict "Sympetalae". Note also that the Loasaceae have the embryological features but not true sympetaly in common with these. In the Santalales, there is an evolutionary depth in ovule characters culminating, in most members, in ategmic ovules with an increasing degree of reduction of the ovule. However, some Olacaceae have uni- or even bitegmic ovules.

Adding a third embryological character, endosperm formation (Fig. 19), one will find that a great many of the sympetalous orders have the cellular type, namely the Corniflorae, Lamiiflorae, Loasiflorae, many Gentianiflorae (except Gentianales), many Solaniflorae and many Asteriflorae. The cellular type, if this is the original state, has broken down in certain lines of evolution within the Sympetalae complex, namely in the Gentianiflorae, where practially all Gentianales have nuclear endosperm formation; in the Solaniflorae, where the cellular type is found in Solanaceae and in those members of Boraginaceae s. lat. which are generally regarded as most "primitive", namely Ehretiaceae and *Heliotropium* of the Boraginaceae s. str. A third order, largely with nuclear endosperm formation, is the Asterales. The nuclear type also is constantly present in Araliiflorae.

On the basis of these characters, the sympetalous orders show some differentiation. Considering these patterns of variation in relation to chemical characters, one will find that most iridoid plants make up one great part of the mostly sympetalous complex with the combination of unitegmic and tenuinnucellate ovules and cellular endosperm formation. Another common characteristic of iridoid plants is the presence of terminal endosperm haustoria. The Gentianales, a group in which the endosperm formation is largely nuclear, is still undoubtedly closely allied to the other orders (with cellular endosperm formation) in this complex, especially the Oleales, Goodeniales and Dipsacales, by virtue of numerous morphological similarities but also the iridoid contents. Also, serological findings support this (J. E. Piechura, pers. com.). Indole and ipecac alkaloids, which are seco-iridoid-derived, are particularly concentrated in the Gentianales, but also found in the Cornales.

In another part of the "Sympetalae", where endosperm formation is partly nuclear, the Asteriflorae, iridoids are totally absent as far as we presently know. Instead, this complex very often possesses polyacetylenes and sesquiterpene lactones. Araliiflorae, which shows embryological similarities to Asteraceae, have in common polyacetylenes as well as sesquiterpene lactones, and the Pittosporaceae have the embryological characters and polyacetylenes.

In the Solanales, there are neither iridoids nor polyacetylenes or sesquiterpene lactones, but the embryological characters and sympetalous conditions are in agreement with the other groups mentioned here, and may be allied to the Asteriflorae or the Lamiiflorae or to both. A richness of other poisonous compounds, such as tropane and pyrrolizidine alkaloids, compensate for the lack of iridoids, polyacetylenes, etc.

In the light of the above, some notes should be made on the taxonomic positions of a few families.

Callitrichaceae, having decussate leaves, flowers situated in the axils of foliose bracts, sometimes capitate glandular hairs and the combination of unitegmic tenuinucellate ovules with cellular endosperm formation and occurrence of endosperm haustoria, is undoubtedly a member of the Sympetalae. Its exact position is perhaps best indicated by the occurrence of schizocarps with one-seeded mericarps and fairly "primitive" iridoids, which in combination indicate a position in the Lamiales (see Dahlgren 1974; Wagenitz 1975).

Hippuridaceae shows similar features, but less specialized embryology. Its embryology is basically of the "Tubiflorean" type and the iridoids primitive. Whether its position is closest to the Cornales or the Scrophulariales is difficult to decide, although a position near Plantaginaceae may be in accord with its general chemical pattern (see Jensen, et al. 1975).

Loasaceae has long been placed in Violales (Parietales) but its embryology has been most aberrant in this order. The ovules are unitegmic and tenuinucellate, the endosperm formation cellular and well-developed endosperm haustoria are present. All this suggests an affinity with the "Sympetalae" complex in spite of the fact that the corolla is choripetalous. The frequently polyandrous androecium is no doubt secondary and derived from a haplostemonous condition. Iridoids are varied and include seco-iridoids, indicating an affinity with orders such as Cornales and Dipsacales. This also was suggested by Kooiman (1974) and Dahlgren (1975).

Sarraceniaceae shows a similar syndrome of characters, combining the same embryological features with ability to form seco-iridoids. Its position near the Cornales-Ericales complex was suggested by Dahlgren (1975), and is more fully explained in Jensen, et al. (1975). Yet another example is the Goodeniaceae which shows morphological, embryological and chemical affinity to the Gentianales (Jensen, et al. 1975).

Other families mentioned by Wagenitz (1975) in some detail are Hydrostachyaceae, Eucommiaceae and Daphniphyllaceae, although the taxonomic positions in these cases are not unambiguous.

The most impressive example of the solution of a similar problem, in which the embryological syndrome of characters mentioned above (but in this case with nuclear endosperm formation) in combination with anatomical, morphological and chemical characters is *Theligonum*. This was, as a rule, placed with the families of Caryophyllales until Wunderlich (1971) revealed its great concordance with Rubiaceae (Gentianales), in which it now has a satisfactory position. This was supported by Kooiman (1971), who detected iridoids in *Theligonum*.

These examples may demonstrate how morphological, anatomical, embryological, and other characters in combination with chemical contents contribute in indicating affinities of families with reduced or specialized vegetative and/or floral morphology.

ACKNOWLEDGEMENTS

In this article, the senior author (Dahlgren) is wholly responsible for the classification and taxonomic aspects presented. The chemical data and comments have been prepared by the authors in cooperation. Thanks are due to many colleagues who have contributed base data and ideas. The data on ellagitannins were kindly supplied by Dr. E. C. Bate-Smith; information for glucosinolate plants has been received from Dr. Ettlinger and Dr. L. Bolt-Jørgensen; and important serological information has been communicated by Dr. D. Fairbrothers and his associates, Drs. P. Petersen and J. E. Piechura. Base data on saponins and chelidonic acid are partly from the card index of Dr. R. Hegnauer. The authors wish to express their appreciation to all these persons for their kind assistance.

We also wish to acknowledge stimulating contacts with Dr. R. F. Thorne, who has, for example, kindly communicated his and Dr. R. Scogin's findings that Simmonsiaceae show close serological affinity to taxa of Euphorbiaceae.

We wish to express our gratitude to the Botanical Society of America, the chairman of its Phytochemical section, Dr. Dale M. Smith and Drs. David A. Young and David S. Seigler for inviting one of us (Dahlgren) to participate in this Symposium, and to the National Science Foundation (USA) for economic support in connection with the transportation and living costs involved. Finally, I thank Dr. D. A. Young for his editorial help with our manuscript.

LITERATURE CITED

Bate-Smith, E. C. 1962. The phenolic constituents of plants and their taxonomic significance. I. Dicotyledons. J. Linn. Soc. (Bot.) 58:95-173.

------. 1972. Chemistry and phylogeny of the angiosperms. Nature 236:353-354.

------. 1974. "Systematic Distribution of Ellagitannins in Relation to the Phylogeny and Classification of the Angiosperms." In Chemistry in Botanical Classification, edited by G. Bendz and J. Santesson, pp. 93-102. New York: Academic Press.

Behnke, H.-D. 1976a. Die siebelement-plastiden der Caryophyllaceae, eine weitere spezifische form der P-typ plastiden bei Centrospermen. Bot. Jahrb. Syst. 95:327-333.

------. 1976b. Ultrastructure of sieve-element plastids in Caryophyllales (Centrospermae), evidence for the delimitation and classification of the order. Plant Syst. Evol. 126:31-54.

------ and R. Dahlgren. 1976. The distribution of characters within an angiosperm system. 2. Sieve tube elements. Bot. Notiser. 129:287-295.

Bohlmann, F., T. Burkhardt and C. Zdero. 1973. Naturally Occurring Acetylenes. London: Academic Press.

Boulter, D. 1973a. "The Molecular Evolution of Higher Plant cytochrome c." In Chemistry in Evolution and Systematics, edited by T. Swain, pp. 539-552. London: Butterworth.

------. 1973b. "The Use of Amino Acid Sequence Data in the Classification of Higher Plants." In Chemistry in Botanical Classification, edited by G. Bendz and J. Santesson, pp. 211-216. New York: Academic Press.

Briggs, B. G. and L. A. S. Johnson. 1979. Evolution in the Myrtaceae--evidence from inflorescence structure. Proc. Linn. Soc. New S. Wales 102:157-256.

Clifford, H. T. 1970. "Monocotyledon Classification with Special Reference to the Origin of the Grasses (Poaceae). In New Research in Plant Anatomy, edited by N. K. B. Robson, D. F. Cutler and M. Gregory, pp. 25-34. London: Academic Press.

Corner, E. J. H. 1976. The Seeds of Dicotyledons. Cambridge: Cambridge University Press.

Cronquist, A. 1979. "Outline of Classification of the Living Families of Seeds Plants." In How to Know the Seed Plants, pp. 136-139. Dubuque, Iowa: W. C. Brown Co.

Culvenor, C. C. J. 1978. Pyrrolizidine alkaloids--occurrence and systematic importance in angiosperms. Bot. Notiser 131: 473-486.

Dahlgren, R. 1975a. A system of classification of the angiosperms to be used to demonstrate the distribution of characters. Bot. Notiser 128:119-147.

------. 1975b. The distribution of characters within an angiosperm system. I. Some embryological characters. Bot. Notiser. 128:181-197.

------. 1977a. A commentary on a diagrammatic presentation of the angiosperms in relation to the distribution of character states. Plant Syst. Evol., Suppl. 1:253-283.

------. 1977b. A note on the taxonomy of the "Sympetalae" and related groups. Publ. Cairo Univ. Herb. 7-8:83-102.

------. 1979. "Gross-taxonomical Evaluations in the Angiosperms in Relation to Parasitism." In Parasites as Plant Taxonomists. New York (in press).

------ and H. T. Clifford (in prep.). The monocotyledons. A comparative study.

------, S. R. Jensen and B. J. Nielsen. 1976. Iridoid compounds in Fouquieriaceae and notes on its possible affinities. Bot. Notiser 129:207-212.

Fish, F. and P. G. Waterman. 1973. Chemosystematics in the Rutaceae. II. The chemosystematics in the Zanthoxylum/ Fagara complex. Taxon 22:177-203.

Gornall, R. J., B. A. Bohm and R. Dahlgren. 1979. The distribution of flavonoids in the angiosperms. Bot. Notiser 132: 1-30.

Goldblatt, P., J. W. Nowicke, T. J. Mabry and H.-D. Behnke. 1976. Gyrostemonaceae: status and affinity. Bot. Notiser 129:201-206.

Gupta, B. D., S. K. Banerjee and K. L. Handa. 1976. Alkaloids and coumarins of *Heracleum wallichii*. Phytochemistry 15: 576.

Harborne, J. B. 1973. "Flavonoids as Systematic Markers in the Angiosperms." In Chemistry in Botanical Classification, edited by G. Bendz and J. Santesson. New York: Academic Press.

Hegnauer, R. 1964. Chemotaxonomie der Pflanzen. Vol. 3. Basel: Birkhauser-Verlag.

------. 1969. Chemotaxonomie der Pflanzen. Vol. 5. Basel: Birkhauser-Verlag.

------. 1977. "The Chemistry of the Compositae." In The Biology and Chemistry of the Compositae, edited by V. H. Heywood, J. B. Harborne and B. L. Turner, pp. 283-335. New York: Academic Press.

------ and P. Kooiman. 1978. Die systematische bedeutung von iridoiden inhaltsstoffen im rahmen von Wettstein's Tubiflorae. Planta Medica 33:1-33.

Herz, W. 1977. "Sesquiterpene Lactones in the Compositae." In The Biology and Chemistry of the Compositae, edited by V. H. Heywood, J. B. Harborne and B. L. Turner, pp. 337-357. London: Academic Press.

Hickey, L. J. and J. A. Wolfe. 1975. The bases of angiosperm phylogeny: vegetative morphology. Ann. Missouri Bot. Gard. 62:538-589.

Jensen, S. R., B. J. Nielsen and R. Dahlgren. 1975. Iridoid compounds, their occurrence and systematic importance in the angiosperms. Bot. Notiser 128:148-180.

Jensen, U. 1968. Serologische beitrage zur systematik der Ranunculaceae. Bot. Jahrb. Syst. 87:

Kolbe, K.-P. 1978. Serologischer beitrag zur systematik der Capparales. Bot. Jahrb. Syst. 99:468-489.

Kooiman, P. 1971. Ein phytochemischer beitrag zur losung des verwandtschaftsproblems der Theligonaceae. Osterreich. Bot. Zeitschr. 119:395-398.

------. 1974. Iridoid glucosides in the Loasaceae and the taxonomic position of the family. Acta Bot. Neerl. 23:677-679.

Kubitzki, K. 1969. Chemosystematische betrachtungen zur grossgliederung der Dicotylen. Taxon 18:360-368,

Mabry, T. J. 1966. "The Betacyanins and Betaxanthins." In Comparative Phytochemistry, edited by T. Swain, pp. 231-244. New York: Academic Press.

------. 1976. Pigment dichotomy and DNA-RNA hybridization data for Centrospermous families. Plant Syst. Evol. 126:79-94.

------. 1977. The order Centrospermae. Ann. Missouri Bot. Gard. 64:210-220.

Philipson, W. R. 1974. Ovular morphology and the major classification of the dicotyledons. Bot. J. Linn. Soc. 68:89-108.

------. 1975. Evolutionary lines within the dicotyledons. New Zealand J. Bot. 13:73-91.

Romeike, A. 1978. Tropane alkaloids--occurrence and systematic importance in angiosperms. Bot. Notiser 131:85-96.

Seigler, D. S. 1977. Plant systematics and alkaloids. The Alkaloids 16:1-82.

Sorensen, N. A. 1977. "Polyacetylenes and Conservatism of Chemical Characters in the Compositae." In The Biology and Chemistry of the Compositae, edited by V. H. Heywood, J. B. Harborne and B. L. Turner, pp. 385-409. New York: Academic Press.

Thorne, R. F. 1973. The "Amentiferae" or Hamamelidae as an artificial group: a summary statement. Brittonia 25:395-405.

------. 1976. A phylogenetic classification of the Angiospermae. Evol. Biology 9:35-106.

Wagenitz, G. 1975. Blutenreduktion als ein zentrales problem der Angiospermen-systematik. Bot. Jahrb. Syst. 96:448-470.

Wunderlich, R. 1971. Die systematische stellung von Theligonum. Osterriech. Bot. Zeitschr. 119:329-394.

APPENDIX I

A Revised Classification of the Angiosperms

DICOTYLEDONEAE

Magnoliiflorae

    Annonales: Annonaceae, Myristicaceae, Eupomatiaceae, Canellaceae
    Aristolochiales: Aristolochiaceae
    Rafflesiales: Rafflesiaceae (incl. Cytinaceae and Mitrastemonaceae), Hydnoraceae
    Magnoliales: Winteraceae, Degeneriaceae, Himantandraceae, Magnoliaceae, Lactoridaceae, Chloranthaceae
    Illiciales: Illiciaceae, Schisandraceae
    Laurales: Amborellaceae, Austrobaileyaceae, Trimeniaceae, Monomiaceae (incl. Siparunaceae and Atherospermataceae), Gomortegaceae, Calycanthaceae (incl. Idiospermaceae), Lauraceae, Hernandiaceae (incl. Gyrocarpaceae)
    Nelumbonales: Nelumbonaceae

Nymphaeiflorae

    Piperales: Saururaceae, Piperaceae (incl. Peperomiaceae)
    Nymphaeales: Cabombaceae, Ceratophyllaceae, Nymphaeaceae (incl. Barclayaceae)

Ranunculiflorae

    Ranunculales: Lardizabalaceae, Sargentodoxaceae, Menispermaceae, Kingdoniaceae, Circaeasteraceae, Ranunculaceae (incl. Hydrastidaceae), Berberidaceae (incl. Glaucidiaceae, Leonticaceae and Podophyllaceae), Nandinaceae
    Papaverales: Papaveraceae, Fumariaceae (incl. Hypecoaceae)

Caryophylliflorae

    Caryophyllales: Phytolaccaceae (incl. Achatocarpaceae, Agdestidaceae and *Limeum*), Basellaceae, Portulacaceae, Stegnospermataceae, Nyctaginaceae, Aizoaceae (incl. Mesembryanthemaceae and Tetragoniaceae), Didiereaceae, Cactaceae, Hectorellaceae, Halophytaceae, Chenopodiaceae (incl. Dysphaniaceae), Amaranthaceae, Molluginaceae, Caryophyllaceae (incl. Illecebraceae)

Polygoniflorae

    Polygonales: Polygonaceae

Malviflorae (=Dilleniiflorae)

    Paeoniales: Paeoniaceae
    Dilleniales: Dilleniaceae
    Malvales: Sterculiaceae, Elaeocarpaceae, Plagiopteraceae, Bixaceae, Cochlospermaceae, Cistaceae, Sphaerosepalaceae, Sarcolaenaceae, Huaceae (position uncertain), Tiliaceae, Dipterocarpaceae, Malvaceae, Bombaceae
    Urticales: Ulmaceae, Moraceae, Cecropiaceae, Barbeyaceae, Cannabinaceae, Urticaceae
    Rhamnales: Rhamnaceae
    Elaeagnales: Elaeagnaceae
    Plumbaginales (position uncertain): Limoniaceae, Plumbaginaceae
    Thymelaeales: Thymelaeaceae
    Euphorbiales: Euphorbiaceae (incl. Picrodendraceae, Hymenocardiaceae, and Uapacaceae), Simmondsiaceae, Pandaceae, Aextoxicaceae (position uncertain), Dichapetalaceae, Didymelaceae (position uncertain; alternatively in Buxales)

Violiflorae

    Violales: Flacourtiaceae (incl. Lacistemaceae), Passifloraceae, Dipentodontaceae, Peridiscaceae (position uncertain), Scyphostegiaceae, Violaceae, Turneraceae, Malesherbiaceae, Achariaceae, Datiscaceae, Begoniaceae, Cucurbitaceae, Caricaceae
    Salicales: Salicaceae
    Tamaricales: Tamaricaceae, Frankeniaceae
    Capparales: Capparaceae (incl. Cleomaceae, Pentadiplandraceae and Koeberliniaceae), Brassicaceae, Tovariaceae, Resedaceae, Gyrostemonaceae, Batidaceae, Moringaceae (position of the three last families somewhat uncertain but probably justifiable here (L. Bolt-Jorgensen, pers. comm.)
    Salvadorales (position uncertain): Salvadoraceae

Theiflorae

    Theales: Stachyuraceae, Pentaphylacaceae (position uncertain), Marcgraviaceae, Quiinaceae, Ancistrocladaceae, Dioncophyllaceae, Nepenthaceae, Medusagynaceae (position uncertain), Caryocaraceae, Strasburgeriaceae, Ochnaceae, Oncothecaceae, Scytopetalaceae (position uncertain), Lecythidaceae (incl. Asteranthaceae, Foetidiaceae, Barringtoniaceae and Napoleonaceae), Theaceae (incl. Tetrameristaceae and Pellicieraceae), Hypericaceae (=Clusiaceae), Elatinaceae
    Droserales: Droseraceae, Lepuropetalaceae, Parnassiaceae

Primuliflorae

    Ebenales: Ebenaceae, Sapotaceae, Styracaceae, Lissocarpaceae
    Primulales: Myrsinaceae, Aegicerataceae, Theophrastaceae, Primulaceae, Coridaceae

Rosiflorae

    Trochodendrales: Trochodendraceae, Tetracentraceae, Eupteleaceae, Cercidiphyllaceae
    Hamamelidales: Hamamelidaceae (incl. Rhodoleiaceae and Altingiaceae), Platanaceae, Myrothamnaceae, Geissolomataceae
    Fagales: Fagaceae, Corylaceae, Betulaceae
    Balanopales: Balanopaceae
    Juglandales: Rhoipteleaceae, Juglandaceae
    Myricales: Myricaceae
    Casuarinales: Casuarinaceae
    Buxales: Buxaceae, Daphniphyllaceae
    Cunoniales: Cunoniaceae, Baueraceae, Ribesiaceae, Brunelliaceae, Davidsoniaceae, Eucryphiaceae, Bruniaceae, Grubbiaceae
    Saxifragales: Crassulaceae, Cephalotaceae, Iteaceae, Francoaceae, Saxifragaceae, Vahliaceae, Greyiaceae
    Gunnerales: Gunneraceae
    Rosales: Crossosomataceae (position uncertain), Rosaceae, Neuradaceae, Malaceae (=Pomaceae), Amygdalaceae, Chrysobalanaceae

Podostemiflorae

    Podostemales: Podostemaceae (incl. Tristichaceae)

Fabiflorae

    Fabales: Mimosaceae, Caesalpiniaceae, Fabaceae

Proteiflorae

    Proteales: Proteaceae

Myrtiflorae

    Myrtales: Myrtaceae (incl. Heteropyxidaceae), Psiloxylaceae, Oliniaceae, Melastomataceae (incl. Memecylaceae), Penaeaceae, Crypteroniaceae, Lythraceae, Sonneratiaceae, Punicaceae, Combretaceae, Onagraceae, Trapaceae (position uncertain)
    Haloragales: Haloragaceae
    Rhizophorales: Rhizophoraceae (excl. Anisophylloideae)

Rutiflorae

    Rutales: Rutaceae (incl. Rhabdodendraceae and Flindersiaceae), Cneoraceae, Surianaceae, Simaroubaceae, Burseraceae, Meliaceae (incl. Aitoniaceae),
    Sapindales: Coriariaceae, Anacardiaceae (incl. Pistaciaceae and Julianiaceae), Leitneriaceae, Podoceae, Sapindaceae (incl. Stylobasidiaceae), Hippocastanaceae, Aceraceae, Akaniaceae, Bretschneideraceae (position uncertain), Emblingiaceae, Meliosmaceae, Staphyleaceae, Sabiaceae (position uncertain), Connaraceae, Melianthaceae (position uncertain)
    Balsaminales: Balsaminaceae
    Polygalaes: Malpighiaceae, Trigoniaceae, Vochysiaceae, Polygalaceae (incl. Xanthophyllaceae and Diclidantheraceae), Krameriaceae
    Geraniales: Zygophyllaceae, Nitrariaceae, Peganaceae, Balanitaceae, Erythroxylaceae, Houmiriaceae, Linaceae, Ctenolophaceae, Ixonanthaceae, Lepidobotryaceae, Oxalidaceae (incl. Averrhoaceae), Geraniaceae, Dirachmaceae, Ledocarpaceae, Vivianiaceae, Biebersteiniaceae
    Tropaeolales (position uncertain): Tropaeolaceae, Limnanthaceae

Santaliflorae

    Celastrales: Celastraceae (incl. Hippocrateaceae, Tripterygiaceae, Siphonodontaceae and Goupiaceae), Stackhousiaceae, Lophopyxidaceae, Cardiopteridaceae (position uncertain), Corynocarpaceae (position uncertain)
    Vitales: Vitaceae (incl. Leeaceae)
    Santalales: Olacaceae (incl. Octoknemataceae), Opiliaceae, Loranthaceae, Misodendraceae, Eremolepidaceae, Santalaceae, Viscaceae

Balanophoriflorae

    Balanophorales (position uncertain): Cynomoriaceae, Balanophoraceae

Araliiflorae

    Pittosporales: Pittosporaceae, Tremandraceae (position uncertain), Byblidaceae
    Araliales: Torricelliaceae, Araliaceae, Apiaceae

Asteriflorae

    Campanulales: Pentaphragmataceae, Campanulaceae (incl. Sphenocleaceae), Lobeliaceae
    Asterales: Asteraceae

Solaniflorae

    Solanales: Solanaceae (incl. Nolanaceae), Duckeodendraceae, Sclerophylacaceae, Goetzeaceae (position uncertain), Convolvulaceae (incl. Humbertiaceae), Cuscutaceae, Cobaeaceae, Polemoniaceae
    Boraginales: Hydrophyllaceae, Ehretiaceae, Boraginaceae, Wellstediaceae, Lennoaceae (position uncertain), Hoplestigmataceae (position uncertain)

Corniflorae

    Fouquierales: Fouquieriaceae

Ericales: Actinidiaceae (incl. Saurauiaceae), Clethraceae, Cyrillaceae, Ericaceae, Empetraceae, Monotropaceae, Pyrolaceae, Epacridaceae, Roridulaceae, Diapensiaceae
Eucommiales: Eucommiaceae
Sarraceniales: Sarraceniaceae
Cornales: Garryaceae, Alangiaceae, Nyssaceae, Cornaceae, Davidiaceae, Helwingiaceae, Phellinaceae, Aquifoliaceae, Paracryphiaceae, Sphenostemonaceae, Symplocaceae, Anisophyllaceae (position uncertain), Icacinaceae, Escalloniaceae, Montiniaceae, Medusandraceae (position uncertain), Columelliaceae, Stylidiaceae (incl. Donatiaceae), Alseuosmiaceae, Hydrangeaceae, Dialypetalanthaceae, Sambucaceae, Adoxaceae, Dulongiaceae, Tribelaceae, Eremosynaceae, Pterostemonaceae, Tetracarpaeaceae (position of the five last families uncertain)
Dipsacales: Caprifoliaceae, Viburnaceae, Valerianaceae, Triplostegiaceae, Dipsacaceae, Morinaceae, Calyceraceae

Loasiflorae

Loasales: Loasaceae

Gentianiflorae

Goodeniales: Goodeniaceae (incl. Brunoniaceae)
Oleales: Oleaceae
Gentianales: Loganiaceae (incl. Antoniaceae, Spigeliaceae, Strychnaceae, Potaliaceae), Rubiaceae (incl. Theligonaceae), Menyanthaceae, Gentianaceae, Apocynaceae, Asclepiadaceae

Lamiiflorae

Scrophulariales: Bignoniaceae, Myoporaceae, Gesneriaceae, Buddlejaceae, Scrophulariaceae (incl. Nelsoniaceae and Orobanchaceae), Globulariaceae, Selaginaceae, Stilbaceae, Retziaceae, Plantaginaceae, Lentibulariaceae, Pedaliaceae, Trapellaceae, Martyniaceae, Acanthaceae, Thunbergiaceae, Mendonciaceae, Henriqueziaceae (position uncertain)
Hippuridales: Hippuridaceae
Hydrostachyales (position uncertain, possibly allied to Cunoniales or Gunnerales in the Rosiflorae): Hydrostachyaceae
Lamiales: Verbenaceae (incl. Phrymaceae but excl. Stilbaceae), Callitrichaceae, Lamiaceae

MONOCOTYLEDONEAE

Alismatiflorae

Hydrocharitales: Butomaceae, Aponogetonaceae, Hydrocharitaceae (incl. Thalassiaceae and Halophilaceae)
Alismatales: Alismataceae (incl. Limnocharitaceae)
Zosterales: Scheuchzeriaceae, Juncaginaceae (incl. Lilaeaceae), Najadaceae, Potamogetonaceae (incl. Ruppiaceae), Zosteraceae, Posidoniaceae, Cymodoceaceae, Zannichelliaceae

Triuridiflorae

Triuridales: Triuridaceae

Ariflorae

Arales: Araceae, Lemnaceae

Liliiflorae

Dioscoreales: Dioscoreaceae (incl. Stenomeridaceae), Trichopodaceae, Taccaceae, Stemonaceae (incl. Croomiaceae), Trilliaceae
Asparagales: Philesiaceae, Luzuriagaceae, Geitonoplesiaceae, Smilacaceae (incl. Rhipogonaceae), Petermanniaceae, Convallariaceae, Asparagaceae, Ruscaceae, Herreriaceae, Dracaenaceae, Nolinaceae, Doryanthaceae, Dasypogonaceae, Hanguanaceae (position uncertain), Xanthorrhoeaceae, Agavaceae, Hypoxidaceae, Tecophilaeaceae, Cyanastraceae, Phormiaceae, Dianellaceae, Eriospermaceae, Asteliaceae, Aphyllanthaceae, Anthericaceae, Asphodelaceae (incl. Aloeaceae), Hemerocallidaceae, Funkiaceae, Hyacinthaceae, Alliaceae (incl. Agapanthaceae and Gilliesiaceae), Amaryllidaceae
Hydatellales (position uncertain): Hydatellaceae
Liliales: Colchicaceae, Iridaceae, Geosiridaceae, Calochortaceae, Tricyrtidaceae, Alstroemeriaceae, Liliaceae, Melanthiaceae
Burmanniales: Burmanniaceae, Thismiaceae, Corsiaceae
Orchidales: Apostasiaceae, Cypripediaceae, Orchidaceae
Velloziales: Valloziaceae
Bromeliales; Bromeliaceae
Haemodorales: Haemodoraceae (incl. Conostylidaceae)
Pontederiales: Pontederiaceae
Philydrales: Philydraceae
Typhales: Sparganiaceae, Typhaceae

Zingiberiflorae

Zingiberales: Lowiaceae, Musaceae, Heliconiaceae, Strelitziaceae, Zingiberaceae, Costaceae, Cannaceae, Marantaceae

Commeliniflorae

Commelinales: Mayacaceae, Commelinaceae (incl. Cartonemataceae)
Eriocaulales: Rapateaceae, Xyridaceae, Eriocaulaceae
Juncales: Thurniaceae, Juncaceae
Cyperales: Cyperaceae
Restionales: Restionaceae (incl. Anarthriaceae and Ecdeiocoleaceae), Centrolepidaceae
Poales: Flagellariaceae (position uncertain), Joinvilleaceae, Poaceae

Areciflorae

Arecales: Arecaceae
Cyclanthales: Cyclanthaceae
Pandanales: Pandanaceae

THE USEFULNESS OF FLAVONOIDS IN ANGIOSPERM PHYLOGENY: SOME SELECTED EXAMPLES

David A. Young

Department of Botany
University of Illinois
Urbana, Illinois 61801

Flavonoid pigments are a group of low molecular weight, biosynthetically related compounds of which nearly 1500 different types have been identified (Hahlbrock and Grisebach 1975, Harborne 1977a). In the past 15 or so years, the usefulness of comparative flavonoid chemistry in plant systematics clearly has been demonstrated. During this time, the majority of studies have been concerned with the application of flavonoid data to taxonomic decisions at the generic level and below (e.g., see review by Crawford 1978), although some (e.g., Harborne 1977a; Giannasi 1978, Gornall, et al. 1979) have advocated their applicability at higher taxonomic categories. Preparing a review of the use of flavonoid data in solving or attempting to resolve systematic problems at higher taxonomic levels within the angiosperms was a difficult task for several reasons. Firstly, many of the more common flavonoids (e.g., quercitin, kaempferol and many of their glycosides) are of such ubiquitous distribution in flowering plants as to be of limited taxonomic value. Secondly, comparatively few studies have been performed which utilize flavonoids at these taxonomic levels. Notable exceptions to this last point are some of the studies of Harborne and his colleagues (see Harborne 1977a and references therein), and the recent publication of Gornall, et al. (1979) on the general distribution of flavonoids within the angiosperms. Lastly, the diversity and structural complexity of flavonoids in flowering plants makes such a review an awesome endeavor. Thus, I have elected to restrict my discussion to three major topics: (1) the distribution of 5- and 7-deoxyflavonoids; (2) the distribution of flavonoid sulfates; and (3)

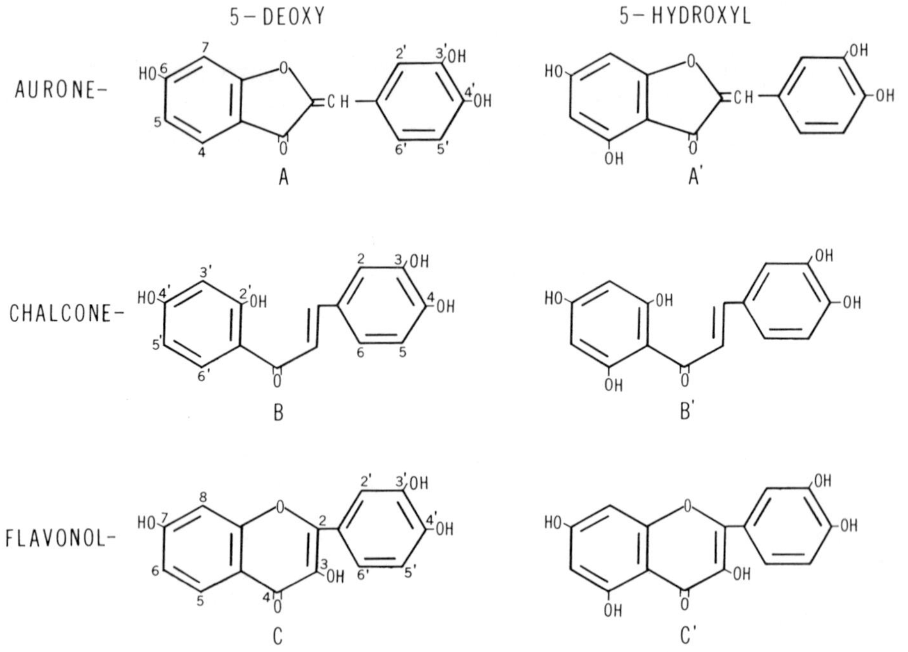

Figure 1. Molecular structures of some typical 5-deoxyflavonoids and their 5-hydroxyl analogs (note differences in the numbering schemes of aurones and chalcones; see text for explanation). A = sulphuretin, A' - dureusidin (rengorsin = 4-methoxy aureusidin); B = butein, B' - 2', 4', 6', 3, 4-pentahydroxy-chalcone (not known as a plant natural product); C = fisetin; C' - quercetin

flavonoids of the Chenopodiiflorae (Centrospermae) and their usefulness in determining the relationships of this superorder. For more general reviews, the reader is referred, for example, to the works of Harborne (1977a), Giannasi (1978) and Crawford (1978). Much of the raw data used in the present discussion was obtained from these reviews, as well as Gornall, et al. (1979) and Harborne, Mabry and Mabry (1975).

I want to stress the point that many of the conclusions presented in this paper may change, perhaps even drastically, as more information on the flavonoids of particular groups of angiosperms (e.g., the Magnoliiflorae) becomes available. Yet this fact should not inhibit our use of currently available data. However, one should remember, as has been pointed out by other participants in this symposium, that single chemical characters --or any characters for that matter--cannot be used by themselves to make taxonomic decisions, but must be used in conjunction with other characters.

5- AND 7-DEOXYFLAVONOIDS

Most classes of flavonoids are characterized by the presence of a hydroxyl group at the 5- and 7- position of the basic flavonoid nucleus (5- and 7-position in flavones, flavonols, flavanones, dihyroflavonols and isoflavones; 4- and 6-position in aurones; 6'- and 4'-position in chalcones). For example, quercitin (3, 5, 7, 3', 4'-pentahydroxyflavone) is a typical 5-hydroxy flavonol, and fisetin (3, 7, 3', 4'-tetrahydroxyflavone) is its 5-deoxy analog. For most of the other classes of flavonoids, there also are 5-deoxy (and often 7-deoxy) analogs of the more common 5-hydroxy compounds (Fig. 1). Although 5-hydroxy and the analogous 5-deoxy compounds have similar R_f values (e.g., R_f of quercitin in TBA/15% HOAc = 0.57/0.03; fisetin = 0.56/0.03), 5-deoxy compounds are easily detected on paper chromatograms since loss of the 5-hydroxyl groups results in intense fluorescence under UV-light. Thus, while quercitin is dull yellow under longwave UV-light, fisetin is bright fluorescent yellow. Because of the difficulties associated with eluting 5-deoxyflavonoids from paper chromatograms, Sephadex column chromatography is the preferred method for the separation and purification of these compounds (see Markham 1975).

Unfortunately, we have few data concerning the biosynthetic pathway(s) leading to the formation of 5- and 7-deoxyflavonoids. However, the data that are available (Rickards 1961) suggest that the oxygen group is removed early in the biosynthet-

Table 1. The distribution of 5- and 7-deoxyflavonoids in angiosperms. (Numbers in parentheses refer to presence of 5- and/or 7-deoxy compounds.)

Dicotyledons:

  Dilleniidae

    Dilleniiflorae: Primulaceae (5, 7)

  Rosidae

    Rutiflorae: Fabaceae (5, 7), Anacardiaceae (incl. Julianiaceae) (5, 7), Rutaceae (5, 7)
    Geraniiflorae: Geraniaceae (5), Vochysiaceae (5)
    Santaliflorae: Celastraceae (5)
    Myrtiflorae: Combretaceae (5)

  Asteridae

    Asteriflorae: Asteraceae (5, 7)

ic pathway, probably before chalcone formation. Such a scheme was proposed for the biosynthetic relationships of the 5- and 7-deoxyflavonoids in the heartwood of several species of *Rhus* (Anacardiaceae; Young 1979a). Biosynthesis of these flavonoids, as well as flavonoids in general, is an area where much more research is needed.

5- and 7-deoxyflavonoids have been reported from nine families of dicotyledons and two families of monocotyledons (Table 1). Because of their apparent limited distribution, when detected these compounds may be of value in indicating taxonomic relationships. For example, the occurrence of 5-deoxyflavonoids in the Juncaceae and Poaceae could be added to the other numerous chemical (e.g., flavonoid sulfates; Harborne 1975a) and morphological features linking these two families (e.g., see Dahlgren 1980, Takhtajan 1980).

Another case where 5-deoxyflavonoids were taxonomically useful was in determining the affinities of the Julianiaceae (Young 1976). The phylogenetic relationships of this family had been a source of controversy for nearly 100 years. Because members of the family are wind pollinated and exhibit many of the features characteristic of this pollination system (e.g., apetaly, male flowers in loose catkins), the family historically had been included in the Amentiferae (see Stern 1952, 1973), close to the Juglandaceae. Others (e.g., Cronquist 1968, Takhtajan 1980, Thorne 1976) have considered the Julianiaceae to be closely allied with or part of the Anacardiaceae. This latter view was supported by the wood anatomy studies of Stern (1952) and Youngs (1955).

Of the 16 flavonoids detected in the leaves and heartwood of *Amphipterygium adstrigens* (Schlect.) Schiede, seven were 5-deoxyflavonoids, and these were primarily restricted to heartwood tissues (Young 1976). All seven of the 5-deoxyflavonoids also have been reported from various species of Anacardiaceae (Hillis and Inoue 1966, Young 1979a) and some of these compounds (e.g., sulphuretin, fisetin, fustin) have been considered characteristic of the Anacardiaceae as a whole (Hegnauer 1964). In addition, rengasin (3', 4', 6-trihydroxy, 4-methoxyaurone) is known only from species of Anacardiaceae and *A. adstrigens*. None of these compounds was detected in taxa of the Juglandaceae or other families presumed to be closely related to the Julianiaceae that were surveyed. These flavonoid data, together with the close morphological similarities of *Amphipterygium* to many genera of the tribe Rhoeae (particularly *Rhus* and *Pistacia*) convinced me that *Amphipterygium* and *Orthopterygium* should be included in the Anacardiaceae as a subtribe (Julianiinae) of the Rhoeae. Although there are those who do not agree with my taxonomic treatment, this certainly is an example where flavonoids clearly have established the relationships of a higher taxonomic group. (I was pleased to hear that serological data presented at these meetings (Petersen and Fairbrothers 1979) support the conclusions primarily based upon flavonoid chemistry.)

When the overall distribution of 5- and 7-deoxyflavonoids is considered (Table 1), an interesting pattern emerges: the compounds primarily are found in families of the Rosidae (sensu Takhtajan 1980). In particular, there are very close similarities between the 5- and 7-deoxyflavonoids of the Fabaceae and Anacardiaceae (see Harborne, Mabry and Mabry 1975; Young 1979a). This pattern perhaps could be used as another piece of evidence supporting the ideas of Thorne (1981) that the Fabaceae are more closely allied to the Rutales than they are to the Rosales. The presence of 5-deoxyflavonoids in the Geraniaceae and Vochysiaceae (Lopes, et al. 1979) might also support including the Geraniiflorae (sensu Thorne 1981) in the Rosidae or near the Rutales as suggested by Cronquist (1968), Takhtajan (1980) and Dahlgren (1980), rather than in the dilleniaceous complex (see Thorne 1981) where 5- and 7-deoxyflavonoids are known only from the Primulaceae (Gornall, et al. 1979). Flavonoid data and information on stamen maturation (i.e., centrifugal vs. centripetal) for species of *Vantanea* and other Houmiriaceae certainly would be useful in helping to determine the relationships of the Geraniiflorae.

There probably is little doubt that the ability to synthesize 5- and 7-deoxyflavonoids (or most flavonoid types, for that matter) has arose independently in at least several evolutionary lines within the angiosperms, and this must be taken into consideration before taxonomic conclusions, based upon their distributions, are made. Needless to say, this is true of all characters used in systematics. However, it is possible that as more plants are surveyed and we have a better idea of the limits of the distribution of these compounds and the biosynthetic pathways leading to them, that 5- and 7-deoxyflavonoids may prove to be exceedingly useful as indicators of phylogenetic relationships in flowering plants.

Table 2. Known flavonol and flavone sulfates (see Harborne 1977b for plant sources).

Flavonols:

Myricetin: 3-rhamnoside sulfate.
Quercitin: 3-sulfate; 3, 7, 3', 4'-tetrasulfate; 7-sulfate-3 glucuronide; 3-rutinoside sulfate; 3-rhamnoside sulfate.
Kaempferol: 7-sulfate; 7-sulfate-3-glucuronide; 3-rhamnoside sulfate; 7, 4'-dimethylether-3-sulfate; 3, 7-disulfate.
Isorhamnetin: 3-sulfate; 7-sulfate; 7-sulfate-3-glucuronide; 3-rutinoside sulfate.
Rhamnocitrin: 3-sulfate.
Patuletin: 7-sulfate; 7-sulfate-3-glucoside.
Gossypin: 3-sulfate.
Rhamnetin: 3-sulfate; 3, 5, 4'-trisulfate-3'-glucuronide.
Rhamnazin: 3-sulfate.
Tamarixetin: 3-sulfate.

Flavones:

Apigenin: 7-sulfate; 7-glucoside sulfate.
Luteolin: 7-sulfate; 3'-sulfate; 4'-sulfate; 7, 3'-disulfate; 7-sulfate-3'-glusoside; 7-sulfate-3'-retinoside; 7-glucoside sulfate; 7-rutmoside sulfate.
Disometin: 7-sulfate; 7, 3'-disulfate.
Chrysoeriol: 7-sulfate; 7-glucoside sulfate.
Tricetin: 3'-sulfate; 7, 3'-disulfate.
Tricin: 7-glucoside sulfate.
Hypolaetin: 8-sulfate; 8-glucoside-3'-sulfate.
Vitexin: 7-sulfate; 7-glucoside sulfate; 7-rutmoside sulfate.
Isovitexin: 7-sulfate.
Orientin: 7-sulfate; 7-glucoside sulfate.
Iso-orientin: 7-sulfate.

Table 3. The distribution of flavonid sulfates in the angiosperms (also see Table 6).

Dicotyledons:

  Dilleniidae

      Dilleniiflorae: Hypericaceae, Plumbaginaceae, Polygonaceae (also sulfated anthraquinones; see Harborne and Mokhtari 1977).
      Chenopodiiflorae: Chenopodiaceae (also sulfated betalains in Phytolaccaceae and Chenopodiaceae; Harborne 1977b).
      Malviflorae: Bixaceae, Cistaceae, Malvaceae (see Nawwar, et al., 1977).
      Violiflorae: Frankeniaceae, Tamaricaceae

  Rosidae

      Rosiflorae: Rosaceae, Davidsoniaceae (see Wilkins and Bohm 1977).
      Myrtiflorae: Combretaceae (also other sulfated phenolics), Onagraceae (J. E. Averett, pers. comm.).

  Asteridae

      Gentianiflorae: Apiaceae
      Asteriflorae: Asteraceae

Monocotyledons:

  Liliidae

      Liliiflorae: Liliaceae

  Alismatidae

      Alismatiflorae: Alismataceae, Hydrocharitaceae, Zannichelliaceae, Zosteraceae

  Commelinidae

      Commeliniflorae: Restionaceae, Poaceae, Juncaceae

  Arecidae

      Areciflorae: Arecaceae

FLAVONOID SULFATES

Although flavonoid sulfates were first reported from the angiosperms over 40 years ago (Kawaquchi and Kim 1937), it only has been recently that these compounds have been shown to occur rather commonly in flowering plants (Harborne 1975a, 1977b). Most of the recent reports of flavonoid sulfates have come from the studies of Harborne and his colleagues (see e.g., Harborne 1975b, Harborne and Williams 1976, Williams and Harborne 1975, 1977, Williams, et al. 1973). To date, approximately 43 flavonol and flavone sulfates have been reported from over 200 species and approximately 24 families of flowering plants (Tables 2 and 3). (Although flavonoid sulfates have not been reported from ferns, sulfate esters of simple phenolic acids are known; Cooper-Driver and Swain 1975. To my knowledge, flavonoid sulfates have not been reported from gymnosperms.)

Once again, little is known of the biosynthesis of flavonoid sulfates, although Harborne (1977b) has suggested a possible biosynthetic pathway. Among the flavonoid sulfates that have been characterized fully, the sulfate radical has been found to be conjugated either directly to the flavonoid nucleus or to a glycosyl group attached to the flavonoid. Detailed reviews of this class of flavonoids, including a discussion of their detection and identification, has been presented by Harborne (1975a, 1977b).

Based upon the early reports of the distribution of flavonoid sulfates in flowering plants, Harborne and his co-workers (Harborne 1975a) noted that these compounds tended to occur in families that are predominately herbaceous or otherwise advanced morphologically. However, they recognized at the time that this may only have been a reflection of inadequate sampling of woody plants (which now appears to be the case).

Flavonoid Sulfates in Dicotyledons

When the known distribution of flavonoid sulfates is plotted on a system of classification, similar to that of Takhtajan (1980), an interesting pattern is seen. Flavonoid sulfates predominately occur in families of the Dilleniidae (including the Chenopodiiflorae) (Table 3), a fact noted in part by Harborne (1975b, 1977b). In this respect, it may not be fortuitous that many of these same families also synthesize glucosinolates (see Rodman 1981) and/or sulfated cyanogens

Table 4. Distribution of Flavonoid Types in the Violiflorae

Flavonoid type (see table 8)	Flacourtiaceae	Violaceae	CISTACEAE	Turneraceae	Passifloraceae	Caricaceae	Cucurbitaceae	Begoniaceae	Datiscaceae	Loasaceae	Salicaceae	Tamaricaceae	Frankiniaceae	Capparaceae	Resedaceae	Brassicaceae	Moringaceae
A	+	+	+	+	+	+		+			+	+	+				
B		+			+			+			+	+	+		+		
C																	
D		+	+								+						
E	+	+	+	+	+	+	+	+	+	+	+	+	+	+	+	+	+
F		+	+					+	+		+	+	+		+	+	+
G																	
H																	
I																	
J			+									+	+				
K																	
L	+	+	+	+			+				+		+		+	+	
M	+	+					+				+		+			+	
N																	
O																	
P																	
Q		+		+			+	+							+		
R								+							+		
S								+									
T								+								+	

Table 5. Distribution of Flavonoid Types in the Malviflorae

Flavonoid Type (see Table 8)	Sterculiaceae	Elaeocarpaceae	Tiliaceae	Dipterocarpaceae	Sarcolaenaceae	Bombacaceae	Bixaceae	Cochlospermaceae	CISTACEAE	Malvaceae	Ulmaceae	Cannabaceae	Urticaceae	Moraceae	Rhamnaceae	Elaeagnaceae	Euphorbiaceae	Thymelaeaceae
A	+	+	+	+	+	+	+	+	+	+	+	+	+	+	+	+	+	+
B	+	+				+				+				+	+	+	+	
C																		
D		+	+	+	+			+	+	+	+			+			+	
E	+	+	+	+	+	+		+	+	+	+	+	+	+	+	+	+	+
F	+	+	+					+	+						+	+	+	
G	+								+							+		
H																		
I									+									
J									+	+								
K							+											
L	+		+					+	+	+		+	+	+			+	+
M		+								+		+	+	+				+
N																		
O							+			+								
P	+																	
Q		+								+	+	+					+	+
R								+			+		+			+		+
S										+		+		+		+		
T			+							+	+			+	+			

215

(Saupe 1981; D. S. Seigler, pers. comm.). Of particular interest in this group is the Cistaceae. Flavonoid sulfates have been reported from *Helianthum squamatum* (L.) Pers., but were not detected in the Flacourtiaceae, Turneraceae or Violaceae (Harborne 1975b). All of these families, including the Cistaceae, were included by Thorne (1976) in his Cistiflorae, and by Cronquist (1968) and Takhtajan (1980) in their Violales. On the other hand, Dahlgren (1980) placed the Cistaceae near the Malvaceae in his Malvales. Based upon overall similarity of flavonoids, the Cistaceae seem to fit better in the Malvales (Malviflorae sensu Thorne) that they do in the Violales (see Table 4 and 5). In the Malvales, for example, they share the common features of flavonoid sulfates and presence of myricetin, which are rare in the Violales. In addition, unlike many members of the Violales (e.g., Caricaceae, Turneraceae, Passifloraceae), members of the Cistaceae do not synthesize cyanogenic compounds (Saupe 1981). In all, these chemical data tend to support placement of the Cistaceae in the Malvales, near the Bixaceae-Cochlospermaceae-Malvaceae complex, which Thorne (1981) now accepts.

Harborne (1975b), noting the absence of flavonoid sulfates and ellagic acid in the Cochlospermaceae versus their presence in *Bixa* (Bixaceae), suggested that these chemical data supported keeping these two groups as separate families. The Cochlospermaceae differ in their flavonoid chemistry from the Bixaceae in several features (Table 5). The Cochlospermaceae, besides lacking flavonoid sulfates, also lack 8-hydroxyflavones and produce flavanones. In general, the Bixaceae are characterized by flavones, whereas the Cochlospermaceae have flavones and flavonols, including myricetin (absent in Bixaceae). These flavonoid differences certainly seem significant enough, when taken together with morphological differences, to treat the two groups as separate families.

Flavonoid Sulfates in Monocotyledons

The distribution of flavonoid sulfates in monocotyledons has been reviewed in detail by Harborne, and have been discussed by Professor Dahlgren in this symposium. Within the monocotyledons, flavonoid sulfates appear to be concentrated in families of the Alismatiflorae and Commeliniflorae (sensu Thorne 1981). In particular, they mainly have been found in the Poaceae, Juncaceae (Commeliniflorae) and Arecaceae (=Palmae; Areciflorae), and appear to be most diverse in the latter.

Table 6. Taxa surveyed and the occurrence of sulfated flavonoids and other phenolics.

Sulfated Compounds Detected	Sulfated Compounds Absent

Combretaceae:
   Laguncularia racemosa
   Conocarpus erecta
   Lumnitzera racemosa
   Bucida buceras
   Combretum grandiflorum
Malvaceae:
   Paeonia spicata
Apocynaceae:
   Rhabdadenia sp.
	Rhizophoraceae:
	   Rhizophora mangle
	Verbenaceae:
	   Avicenna germinans
	Plumbaginaceae:
	   Aegialitis
	Myrtaceae:
	   Osbornia

Ecological Considerations

Harborne and his associates noted that, based upon the taxa they had surveyed, there was a strong correlation between plants which synthesize flavonoid sulfates and their occurrence in either fresh or salt water or saline habitats, such as salt marshes and alkaline desert areas. For example, all of the flavonoids produced by *Zostera marina* L. are sulfated (Harborne and Williams 1976). However, these conclusions primarily were based upon data from monocotyledons. To test whether or not this correlation was generally applicable to all flowering plants, several mangrove taxa were surveyed for flavonoid sulfates (Young 1979b). One of the more striking associations between dicotyledons and a saline habitat can be seen in mangrove swamps. Ecologically, mangrove swamps represent the tropical equivalent of salt marshes (Chapman 1970). Physiognomically, mangrove swamps differ from salt marshes in that the dominant species are trees rather than herbs. Thus, by looking at mangrove taxa, we were able to test the hypotheses of an ecological factor affecting accumulation of flavonoid sulfates and that they tend to occur primarily in herbaceous groups.

The taxa surveyed and the results of the survey are shown in Table 6. As can be seen, flavonoid sulfates were not detected in *Rhizophora mangle* L. or *Avicenna germinans* L., two of the more common species of mangrove swamps in the new world. However, several species of Combretaceae, including mangrove (e.g., *Laguncularia racemosa* Gertn. and *Conocarpus erecta* L.) and non-mangrove (e.g., *Combretum grandiflorum*) taxa produced sulfated flavonoids (as well as other sulfated phenolics). These data, although limited, suggest that synthesis of flavonoid sulfates may be correlated as much with taxonomic relationships as with ecological factors. Thus, it may be significant that flavonoid sulfates were detected in members of the Myrtales (although absent from *Osbornia*-Myrtaceae), which in general is characterized by the presence of ellagic acid (Hegnauer 1964). The association of ellagic acid with flavonoid sulfates already has been noted by Harborne (1977b). These results also show that flavonoid sulfates are not limited to herbaceous taxa, and the generalization that they are characteristic of herbaceous groups was in fact simply indicative of inadequate sampling of woody plants. Further generalizations must await collection of more data, and given the ease with which flavonoid sulfates can be detected, such tests routinely should be incorporated into all flavonoid surveys. The limited data that we do have certainly suggest that flavonoid sulfates may be of value in

Table 7. Comparison of the classification systems of Cronquist (1968), Dahlgren (1980), Takhtajan (1980) and Thorne (1981) for the Chenopodiiflorae.

Cronquist

Caryophyllidae
    Caryophyllales: Phytolaccaceae, Achatocarpaceae, Nyctaginaceae, Aizoaceae, Didiereaceae, Cactaceae, Chenopodiaceae, Amaranthaceae, Portulacaceae, Basellaceae, Molluginaceae, Caryophyllaceae.
    Polygonales: Polygonaceae.
    Plumbaginales: Plumbaginaceae.

Dahlgren

Caryophylliflorae
    Caryophyllales: Phytolaccaceae, Basellaceae, Portulacaceae Stegnospermataceae, Nyctaginaceae, Aizoaceae, Didiereaceae, Cactaceae, Hectorellaceae, Halophytaceae, Chenopodiaceae, Amaranthaceae, Molluginaceae, Caryophyllaceae.

Takhtajan

Caryophyllidae
    Caryophyllanae
        Caryophyllales: Phytolaccaceae, Achatocarpaceae, Nyctaginaceae, Aizoaceae, Cactaceae, Portulacaceae, Hectorellaceae, Basellaceae, Didiereaceae, Stegnospermataceae, Molluginaceae, Caryophyllaceae, Amaranthaceae, Chenopodiaceae.
    Polygonales: Polygonaceae.
    Plumbaginales: Plumbaginaceae.

Thorne

Chenopodiiflorae
    Chenopodiales: Phytolaccaceae, Aizoaceae (including Molluginaceae), Caryophyllaceae, Halophytaceae, Nyctaginaceae, Chenopodiaceae, Amaranthaceae, Portulacaceae, Basellaceae, Didiereaceae, Cactaceae.

Table 8. Key to the classes of flavonoids used in the analysis of relationships of the Chenopodiiflorae (see Gornall, et al. 1979).

A. Proanthocyanidin
B. Anthocyanidins
C. Betalains
D. Myricetin
E. Quercetin and/or kaempferol
F. Methylated flavonols
G. 8-hydroxy flavonols
H. 6-hydroxy flavonols
I. 5-methoxy flavonols
J. Flavonol sulfates
K. Flavone sulfates
L. Luteolin and/or apigenin
M. Methylated flavones
N. 6- or 8-methoxy flavones
O. 8-hyroxy flavones
P. 6-hydroxy flavones
Q. C-glycosylflavones
R. Flavonones
S. Chalcones
T. Dihydroflavonols

Table 9. Distribution of Flavonoid Types in the Chenopodiiflorae

Flavonoid Type (see Table 8)	Phytolaccaceae	Nyctaginaceae	Aizoaceae	Chenopodiaceae	Halophytaceae and Hectorellaceae	Amaranthaceae	Portulacaceae	Basellaceae	Cactaceae	Didiereaceae	Molluginaceae	Caryophyllaceae
A	+	+					+					+
B											+	+
C	+	+	+	+	+	+	+	+	+	+		
D		+	+							+		
E	+	+	+	+		+	+	+	+	+	+	+
F	+		+	+				+				+
G												
H			+	+								
I												
J			+									
K												
L											+	
M				+								
N						+						
O												
P						+						
Q	+	+		+		+	+	+		+	+	+
R												
S												
T											+	

determining taxonomic relationships.

FLAVONOIDS AND TAXONOMIC RELATIONSHIPS OF THE CHENOPODIIFLORAE

The last topic that I will discuss is the flavonoids of the Chenopodiiflorae (sensu Thorne 1981). The Chenopodiiflorae (=Centrospermae) consist of a single order (Chenopodiales), 13 families and approximately 6700 species. The Chenopodiiflorae generally are recognized as a natural group of closely related families (Table 7) that possess a set of features that readily distinguish them from all other superorders of dicotyledons (see Eckhardt 1976). Some of these features are: (1) free-central or basal placentation; (2) flowers basically with a uniseriate perianth; (3) often numerous, centrifugal stamens; (4) typically bitegmic, crassinucellate ovules; (5) seeds with a curved or coiled embryo and abundant perisperm (little or no endosperm); (6) trinucleate pollen; (7) betalain pigments (except Molluginaceae and Caryophyllaceae); and (8) P-III subtype sieve member plastids. Although much work remains to be done, probably no other superorder of flowering plants of its size has been as well investigated morphologically, ultrastructurally and chemically as the Chenopodiiflorae (Mabry 1977). Yet the phylogenetic relationships of the Chenopodiiflorae remain obscure, and as we heard in this symposium (Scogin 1981) cytochrome $\underline{c}$ and ferredoxin data suggest that the group may be rather isolated from the rest of the angiosperms. Some authors (Buxbaum 1961; Cronquist 1968; Takhtajan 1973, 1980) have suggested that the Chenopodiiflorae may have been derived from the Magnoliiflorae (Magnoliales or Berberidales). Others (Eckhardt 1976; Thorne 1976) cite apocarpy and centrifugal development of stamens in the Chenopodiiflorae as indicating a relationship with the Dilleniidae (particularly Dilleniiflorae-Dilleniales). The purpose of this discussion is to compare the flavonoids of the Chenopodiiflorae with those of the Magnoliidae and Dilleniidae to determine whether or not they indicate or suggest a relationship with either of the latter two groups. The flavonoid types utilized are listed in Table 8, and the data were obtained primarily from Gornall, et al. (1979) and Richardson (1978).

Flavonoids of the Chenopodiiflorae

The distribution of selected flavonoid types in the Chenopodiiflorae is presented in Table 9. Besides betalains (non-flavonoids), the group in general is characterized by the pres-

Table 10. Distribution of Flavonoid Types in the Magnoliiflorae

Flavonoid Type (see Table 8)	Winteraceae	Illiciaceae	Schisandraceae	Magnoliaceae	Degeneriaceae	Eupomatiaceae	Annonaceae	Myristicaceae	Aristolochiaceae	Chloranthaceae	Monimiaceae	Lauraceae	Calycanthaceae	Idiospermaceae	Hernandiaceae	Saururaceae	Piperaceae	Nelumbonaceae	Lardizabalaceae	Menispermaceae	Berberidaceae	Ranunculaceae	Papaveraceae
A	+	+	+	+		+	+			+	+					+	+	+			+	+	
B			+	+					+	+	+	+							+		+	+	+
C																							
D				+																	+		
E	+	+	+	+	+		+	+		+	+	+		+	+		+	+	+		+	+	+
F				+					+	+	+										+	+	+
G																						+	+
H																							
I											+												
J																							
K																							
L	+					+							+				+					+	
M																			+		+		
N								+															
O								+															
P								+															
Q								+				+			+							+	
R						+	+	+							+								
S								+							+								
T	+							+															

ence of flavonols and methylated flavonols, and C-glycosylflavones. Flavones are rare. Proanthocyanidins occur in four families (Phytolaccaceae, Nyctaginaceae, Portulacaceae and Caryophyllaceae), and myricetin in three (Aizoaceae, Chenopodiaceae and Didiereaeceae. Other types, including flavonoid sulfates (Chenopodiaceae) occur sporadically throughout the superorder.

Some interesting relationships emerge from these data. There appears to be a close relationship between the flavonoids of the Aizoaceae and Chenopodiaceae. Both are characterized by the presence of myricetin, 6-hydroxyflavonols and methylated flavonols. The Aizoaceae differ from the Chenopodiaceae in lacking C-glycosylflavones. In addition, the Molluginaceae is quite distinct from the Aizoaceae, with which it is sometimes allied (Thorne 1981). The Molluginaceae lack betalains and myricetin, but have C-glycosylflavones. These data would tend to support separating the Molluginaceae from the Aizoaceae.

Flavonoids of the Magnoliiflorae

Flavonoids for families of the Magnoliiflorae for which we have data are presented in Table 10. Unfortunately, little or no data are available for the Himantandraceae, Eupomatiaceae, Annonaceae, Canellaceae, Amborellaceae, Austrobaileyaceae, Chloranthaceae, Lactoridaceae, Gomortegaceae and Sargentodoxaceae (see Young and Sterner 1981 for a discussion of the flavonoids and phylogenetic relationships of the Degeneriaceae and Idiospermaceae). This lack of data severly limits the conclusions that can be made about the Magnoliiflorae, but based upon the data that are available the group seems to be characterized by proanthocyanidins, common flavonols and methylated flavonols, and the absence of myricetin (except in Schisandraceae and Berberidaceae). Flavanones occur in four families. With respect to flavonoids, the Lauraceae is by far the most diverse family in the superorder, but it also has been studied in much more detail than the other families. The Ranunculaceae and Papaveraceae are distinctive in synthesizing 8-hydroxyflavonols. In general, flavones appear to be rare in the Magnoliiflorae, and flavonoid sulfates have not been reported from the superorder.

Flavonoids of the Dilleniidae

In contrast to the Magnoliiflorae, the flavonoids of many

Table 11. Distribution of Flavonoid Types in the Dilleniiflorae

Flavonoid Type (see Table 8)	Dilleniaceae	Paeoniaceae	Actinidiaceae	Stachyuraceae	Theaceae	Icacinaceae	Aquifoliaceae	Marcgraviaceae	Clethraceae	Cyrillaceae	Sarraceniaceae	Ochnaceae	Nepenthaceae	Hyperiaceae	Elatinaceae	Lecythidaceae	Ericaceae	Epacridaceae	Empetraceae	Ebenaceae	Sapotaceae	Symplocaceae	Styraceae	Primulaceae	Theophrastaceae	Myrsinaceae	Plumbaginaceae	Polygonaceae
A	+		+	+	+	+		+	+	+	+	+	+	+	+	+	+	+	+	+	+	+	+	+		+	+	+
B		+			+		+					+		+	+	+	+							+		+	+	+
C																												
D	+			+			+		+					+			+	+		+	+			+		+	+	+
E	+	+	+	+	+	+	+	+	+	+	+		+	+	+	+	+	+	+	+	+	+	+	+	+	+	+	+
F	+			+					+			+					+				+			+			+	+
G				+													+		+					+				+
H																												+
I	+																+										+	+
J																											+	+
K													+															
L		+											+	+		+								+			+	+
M																												+
N																						+						
O																						+						
P																								+			+	
Q			+									+					+							+			+	+
R																	+											
S		+															+									+		
T	+																+				+							

225

more families of the Dilleniidae have been surveyed. Still, we have little or no data for the following families: Cardiopteridaceae, Aquifoliaceae, Phellinaceae, Oncothecaceae, Sphenostemonaceae, Parachryphaceae, Caryocaraceae, Pentaphylacaceae, Quiinaceae, Syctopetalaceae, Sphaerosepalaceae, Medusagynaceae, Strasburgeriaceae, Dioncophyllaceae and Lissocarpaceae. For ease of comparison, each of the superorders of the Dilleniidae will be discussed separately.

Flavonoids of the Dilleniiflorae

The distribution of flavonoids in the Dilleniiflorae is shown in Table 11. In general, the superorder is characterized by proanthocyanidins, myricetin and common flavonols and methylated flavonols. 8-hydroxyflavonols occur in five families: Theaceae, Ericaceae, Empetraceae, Primulaceae and Polygonaceae (6-hydroxyflavonols also occur in the Polygonaceae). It is perhaps significant that of the 11 families known to synthesize 8-hydroxyflavonols, five are members of the Dilleniiflorae and three (Sterculiaceae, Malvaceae, and Euphorbiaceae) occur in the closely related Malviflorae. 5-methoxyflavonols occur in four families: Dilleniaceae, Ericaceae, Plumbaginaceae and Polygonaceae. Only three other families of flowering plants (Lauraceae, Eucryphiaceae and Juglandaceae; see Harborne 1977a; Gornall, et al. 1979) are known to synthesize 5-methoxyflavonols. Flavones and C-glycosylflavones occur sporadically in the superorder.

Flavonoids of the Malviflorae

The distribution of flavonoids in the Malviflorae is presented in Table 5. The general pattern for the Malviflorae is very similar to that of the Dilleniiflorae (e.g., proanthocyanidins, myricetin, flavonols, 8-hydroxyflavonols; 8-hydroxyflavones in Bixaceae and Malvaceae). However, flavones, methylated flavones and C-glycosylflavones are much more common in the Malviflorae than in the Dilleniiflorae.

Flavonoids in the Violiflorae

Flavonoids in the Violiflorae (less Cistaceae) are shown in Table 4. Unlike the Dilleniiflorae and Malviflorae, proanthocyanidins and myricetin are less common (the latter occurs in only two families), and flavones and C-glycosylflavones are more common in the somewhat more herbaceous Violiflorae.

From the data discussed above, there appears to be a general trend or progression in the flavonoids of the Dilleniidae

Table 12. Proposed classification of the Dilleniidae based upon numerous chemical (including flavonoid) and morphological features.

Subclass: Dilleniidae

    Superorder 1: Dilleniiflorae (Dilleniales; Ericales; Ebenales; Primulales; Plumbaginales; Polygonales)

    Superorder 2: Chenopodiiflorae (Chenopodiales)

    Superorder 3: Malviflorae (Malvales, incl. Cistaceae; Urticales; Rhamnales; Euphorbiales)

    Superorder 4: Violiflorae (Violales; Salicales; Tamaricales; Capparales) that seems to be associated with the shift from a woody to more or less herbaceous habit (i.e., Dilleniiflorae to Malviflorae to Violiflorae). Thus, we see flavonols, myricetin and few flavones and C-glycosylflavones in the Dilleniiflorae; an increase in flavones and c-glycosylflavones in the Malviflorae; and finally myricetin occurring only rarely and an increase in the occurrence of flavones and C-glycosylflavones in the Violiflorae.

When the flavonoid profile of the Chenopodiiflorae is compared with that of the Magnoliiflorae and Dilleniidae, the Chenopodiiflorae seems more similar to the Dilleniiflorae and Malviflorae than it is to the Magnoliflorae. In particular, there is an apparent close relationship between the flavonoids of the Chenopodiiflorae and those of the Primulaceae, Polygonaceae and Plumbaginaceae (Dilleniiflorae) (e.g., myricetin, 6-hydroxyflavonols and flavones, c-glycosylflavones). Of course, there are those who argue (e.g., Cronquist) that the Polygonaceae and Plumbaginaceae should be included in the Chenopodiiflorae (or Caryophyllidae) (see Table 7) and cytochrome c data do link the Polygonaceae and Chenopodiaceae (Boulter 1974). However, these families (i.e., Primulaceae, Polygonaceae and Plumbaginaceae) are anomalous in the Chenopodiiflorae in lacking P-type sieve member plastids and betalains (see Mabry 1977). Whether placed in the Chenopodiiflorae or Dilleniiflorae, the Primulaceae, Polygonaceae and Plumbaginaceae seem to "connect" the two superorders. As mentioned above, there are several morphological features (e.g., centifugal stamens) linking the Chenopodiiflorae to the Dilleniidae, and the flavonoid data, although not overwhelming, certainly do not preclude such a relationship. I suggest that we ought to at least seriously consider the possibility of a close relationship between the Dilleniiflorae and Chenopodiiflorae (see Table 12). It certainly seems to me that recognizing the Chenopodiiflorae as a separate subclass (Caryophyllidae sensu Cronquist and Takhajan) places undue emphasis on rather few characters (e.g., betalain pigments and sieve member plastids), and obscures what appear to be more or less natural affinities between the Chenopodiiflorae and Dilleniidae (i.e., Dilleniiflorae).

To conclude, I am sure that there are some who will argue that I have gone too far or at least stretched the use of flavonoids in assessing phylogenetic relationships, to a point of perhaps losing credibility. Yet my objective has been to point out that much information can be gained from using flavonoids in such a fashion. Certainly, their use should not be restricted to establishing or confirming relationships of lower taxonomic levels. They may be, as Professor Dahlgren might say, "weak" characters, but we have not yet begun to tap their total potential or usefulness in angiosperm phylogeny. However, before this goal can be fully realized, we need considerably more data for many more families. In addition, as others have pointed out (e.g., Gornall and Bohm 1978; Crawford 1978; Giannasi 1978; P. Richardson, manuscript in prep.), hypotheses must be developed and rigorously tested concerning evolutionary trends of flavonoids within the angiosperms. Chemotaxonomists have accomplished a lot in the last two decades, but we still have a long way to go.

ACKNOWLEDGEMENTS

I thank Robert Thorne, Rolf Dahlgren, Art Cronquist, Ron Scogin, Mick Richardson and Robert Patterson for many stimulating discussions concerning flavonoids and angiosperm phylogeny. Of course, I take full responsiblity for the opinions expressed in this paper. Part of this work was supported by NSF grant DEB 78-11183 and grants from the University of Illinois Research Board to the author.

LITERATURE CITED

Boulter, D. 1974. The evolution of plant proteins with special reference to higher plant cytochrome c. Curr. Adv. Plant Sci. 8:1-16.

Buxbaum, F. 1961. Vorläufige untersuchungen über umfang, systematische stellund und gliederung der Caryophyllales (Centrospermae). Beitr. Biol. Pflanzen 36:1-56.

Chapman, V. J. 1970. Mangrove phytosociology. Trop. Ecol. 11:1-19.

Cooper-Driver, G. and T. Swain. 1975. Sulphate esters of caffeyl- and p-coumarylglucose in ferns. Phytochemistry 14:2506-2507.

Crawford, D. J. 1978. Flavonoid chemistry and angiosperm evolution. Bot. Rev. 44:431-456.

Cronquist, A. 1968. The Evolution and Classification of Flowering Plants. Boston: Houghton and Mifflin.

Dahlgren, R. M. T. 1980. A revised system of classification of the angiosperms. Bot. J. Linn. Soc. 80:91-124.

Eckhardt, T. 1976. Classical morphological features of centrospermous families. Plant Syst. Evol. 126:5-25.

Giannasi, D. E. 1978. Systematic aspects of flavonoid biosynthesis and evolution. Bot. Rev. 44:339-429.

Gornall, R. J. and B. A. Bohm. 1978. Angiosperm flavonoid evolution: a reappraisal. Syst. Bot. 3:353-368.

------, ------, and R. Dahlgren. 1979. The distribution of flavonoids in the angiosperms. Bot. Notiser 132:1-30.

Hahlbrock, K. and H. Grisebach. 1975. "Biosynthesis of Flavonoids". In The Flavonoids, edited by J. B. Harborne, et al. London: Champan and Hall.

Harborne, J. B. 1975a. Flavonoid sulphates: a new class of sulphur compounds in higher plants. Phytochemistry 14:1147-1155.

------. 1975b. Flavonoid bisulphates and their co-occurrences with ellagic acid in the Bixaceae, Frankeniaceae and related families. Phytochemistry 14:1331-1337.

------. 1977a. Flavonoids and the evolution of the angiosperms. Biochem. Syst. Ecol. 5:7-22.

------. 1977b. Flavonoid sulphates: a new class of natural product of ecological significance in plants. Prog. Phytochemistry 4:189-208.

------ and N. Mokhtari. 1977. Two sulphated anthraquinone derivatives in *Rumex pulcher*. Phytochemistry 16:1314-1315.

------ and C. A. Williams. 1976. Occurrence of sulphated flavones and caffeic acid esters in members of the Fluviales. Biochem. Syst. Ecol. 4:37-41.

------, T. J. Mabry and H. Mabry (eds.) 1975. The Flavonoids. London: Chapman and Hall.

Hegnauer, R. 1964. Chemotaxonomie der Pflanzen. Stuttgart: Basel.

Hillis, W. E. and T. Inoue. 1966. The formation of polyphenols in trees. III. The effect of enzyme inhibitors. Phytochemistry 5:483-490.

Kawaguchi, R. and K. W. Kim. 1937. J. Pharm. Soc. Japan 57: 108 (cited in Harborne, 1977b).

Lopes, J. L. C., J. N. C. Lopes and H. F. Leitao Filho. 1979. 5-deoxyflavones from the Vochysiaceae. Phytochemistry 18: 362.

Mabry, T. J. 1977. The order Centrospermae. Ann. Missouri Bot. Gard. 64:210-220.

Markham, K. R. 1975. "Isolation Techniques for Flavonids." In The Flavonoids, edited by J. B. Harborne, et al. London: Chapman and Hall.

Nawwar, M. A. M., A. El Dein, A. El Sherbeiny, M. A. El Ansari and H. I. El Sissi. 1977. Two new sulphated flavonol glucosides from leaves of *Malva sylvestris*. Phytochemistry 16:145-146.

Petersen, F. and D. E. Fairbrothers. 1979. *Amphipterygium*-- an amentiferous member of the Anacardiaceae. Phytochem. Bull. 12:28-29. (abstract).

Richardson, M. 1978. Flavonols and C-glycosylflavonoids of the Caryophyllales. Biochem. Syst. Ecol. 6:283-286.

Rodman, J. E. 1981. "Divergence, Convergence, and Parallelism in Phytochemical Characters: the Glucosinolate-Myrosinase System". In Phytochemistry and Angiosperm Phylogeny, edited by D. A. Young and D. S. Seigler. New York: Praeger Scientific.

Saupe, S. 1981. "Cyanogenic Compounds and Angiosperm Phylogeny." In Phytochemistry and Angiosperm Phylogeny, edited by D. A. Young and D. S. Seigler. New York: Praeger Scientific.

Stern, W. L. 1952. The comparative anatomy of the xylem and the phylogeny of the Julianiaceae. Amer. J. Bot. 39:220-229.

------. 1973. Development of the amentiferous concept. Brittonia 25:316-333.

Takhtajan, A. 1973. Evolution und Ausbrietung der Blutenpflanzen. Jena and Stuttgart: G. Fischer.

------. 1980. Outline of the classification of flowering plants (Magnoliophyta). Bot. Rev. 46:225-359.

Thorne, R. F. 1976. "A Phylogenetic Classification of the Angiospermae." In Evolutionary Biology, Vol. 9, edited by M. K. Hecht, et al. New York: Plenum Press.

------. 1981. "Phytochemistry and Angiosperm Phylogeny: A Summary Statement." In Phytochemistry and Angiosperm Phylogeny, edited by D. A. Young and D. S. Seigler. New York: Praeger Scientific.

Wilkins, C. K. and B. A. Bohm. 1977. Flavonoids of *Davidsonia pruriens*. Phytochemistry 16:144-145.

Williams, C. A. and J. B. Harborne. 1975. Luteolin and daphnetin derivatives in the Juncaceae and their systematic significance. Biochem. Syst. Ecol. 3:181-190.

------ and ------. 1977. The leaf flavonoids of the Zingiberaceae. Biochem. Syst. Evol. 5:221-229.

------, ------, and H. T. Clifford. 1973. Negatively charged flavones and tricin as chemosystematic markers in the Palmae. Phytochemistry 12:2417-2430.

Young, D. A. 1976. Flavonoid chemistry and phylogenetic relationships of the Julianiaceae. Syst. Bot. 1:149-162.

------. 1979a. Heartwood flavonoids and the infrageneric relationships of *Rhus* (Anacardiaceae). Amer. J. Bot. 66:502-510.

------. 1979b. The occurrence of flavonoid sulfates in four mangrove species. Phytochem. Bull. 12:27 (abstract).

------ and R. W. Sterner. 1981. Leaf flavonoids of "primitive" dicotyledonous angiosperms: *Degeneria vitiensis* (Degeneriaceae) and *Idiospermum australiense* (Idiospermaceae). Biochem. Syst. Ecol.: in press.

Youngs, R. L. 1955. The xylem anatomy of *Orthopterygium* (Julianiaceae). Trop. Woods 101:29-43.

PHYTOCHEMISTRY AND ANGIOSPERM PHYLOGENY
A SUMMARY STATEMENT

Robert F. Thorne

Rancho Santa Ana
Botanic Gardens
Claremont, California 91711

It would be extremely difficult, especially for a non-chemically trained taxonomist, to summarize the abundant, phyletically valuable, and often rather esoteric information that has been presented in this symposium. Also, it would be superfluous since each contributor has presented a rather tight summary of his own phytochemical subject. Instead, I shall expand upon some of the statements made, while presenting my own high opinion of the value of our newly honed phytochemical tools. I should also like to give some examples of how I have applied the finished products constructed with these tools by others, i.e., the skilled phytochemists, in developing my own peculiar system of angiosperm classification. This system, brought up to date, is represented visually by Figure 1 and by the appended synopsis.

First, I wish to discuss in a general way the application of phytochemical information to the solving of phlylogenetic problems. A most important principle has been stated in this symposium by Rolf Dahlgren (Dahlgren, et al. 1981) that "single chemical properties" can be used for extensive phyletic purposes only when supported by other characteristics. This has also been stated by one of our esteemed fellow phylogenists, Armen Takhtajan of Leningrad, who stated not long ago (1974) that "No one single source of information can substitute for the integrative approach to plant classification based on the correlation and synthesis of the evidence taken from all available sources of knowledge." There are few, if any, panaceas in solving the mysteries of angiosperm origins and relationships. I am extremely suspicious of all those enthusiasts with a new approach, new tool, or new group of chemical compounds who decide they have the ultimate answer to all our phyletic problems. I am also old enough now to suspect that few features, very few, phytochemical or otherwise, are strictly monophyletic. If phytochemists in general were better informed about the prevalence of convergence in complex morphological structures in the angiosperms, they might be considerably less sanguine about the monophyly of their own pet biochemical group. Among some of the best known and most complex of these morphological convergences are the pitchers, or ascidia, of unrelated carnivorous plants (Adams and Smith 1977), sticky glandular hairs of the unrelated fly-trappers (Marburger 1979), succulent stems or leaves, aments of the unrelated anemophilous taxa, wind-blown pollen grains, and the remarkable fruits and seeds (Roth 1977) specialized for dispersal by water currents, wind, birds, bats, or other vectors. Convergence in biochemistry is surely no less prevalent, as illustrated by the common and widespread occurrence of many complex compounds in quite unrelated subclasses or divisions of the plant world (Bate-Smith 1962; Gornall, et al. 1979; Hegnauer 1959, 1960, 1977, 1978a; and Seigler 1977).

To paraphrase Dahlgren again, the distribution of a group of chemical compounds, whether it be glucosinolates, betalains, or monoterpinoid iridoids, is truly phyletically significant when it can be ascertained that that distribution coincides generally with evidence gleaned from careful study of the stem anatomy, floral morphology, leaf structure, embryology, pollen grain, fruit and seed morphology, etc. The differences in the various angiosperm classifications devised by Takhtajan (1959), Cronquist (1968), Hutchinson (1973), Dahlgren (1975) and myself, and others basically comes down to our interpretation of just how much coincidence there is in these approaches. I do not, for example, believe that the iridoids or the glucosinolates are monophyletic, hence restricted to just one closely related group of orders and families. I believe that they have evolved several times, perhaps along different biosynthetic pathways. (I do have to believe that the betalains are monophyletic since they are apparently restricted to one order of dicotyledons, the Chenopodiales.) This creates considerable divergence in our classifications. That is not all bad because it gives specialists more than one target to shoot at. Time ultimately will tell which of us is correct or, what is quite possible, that all of us are wrong.

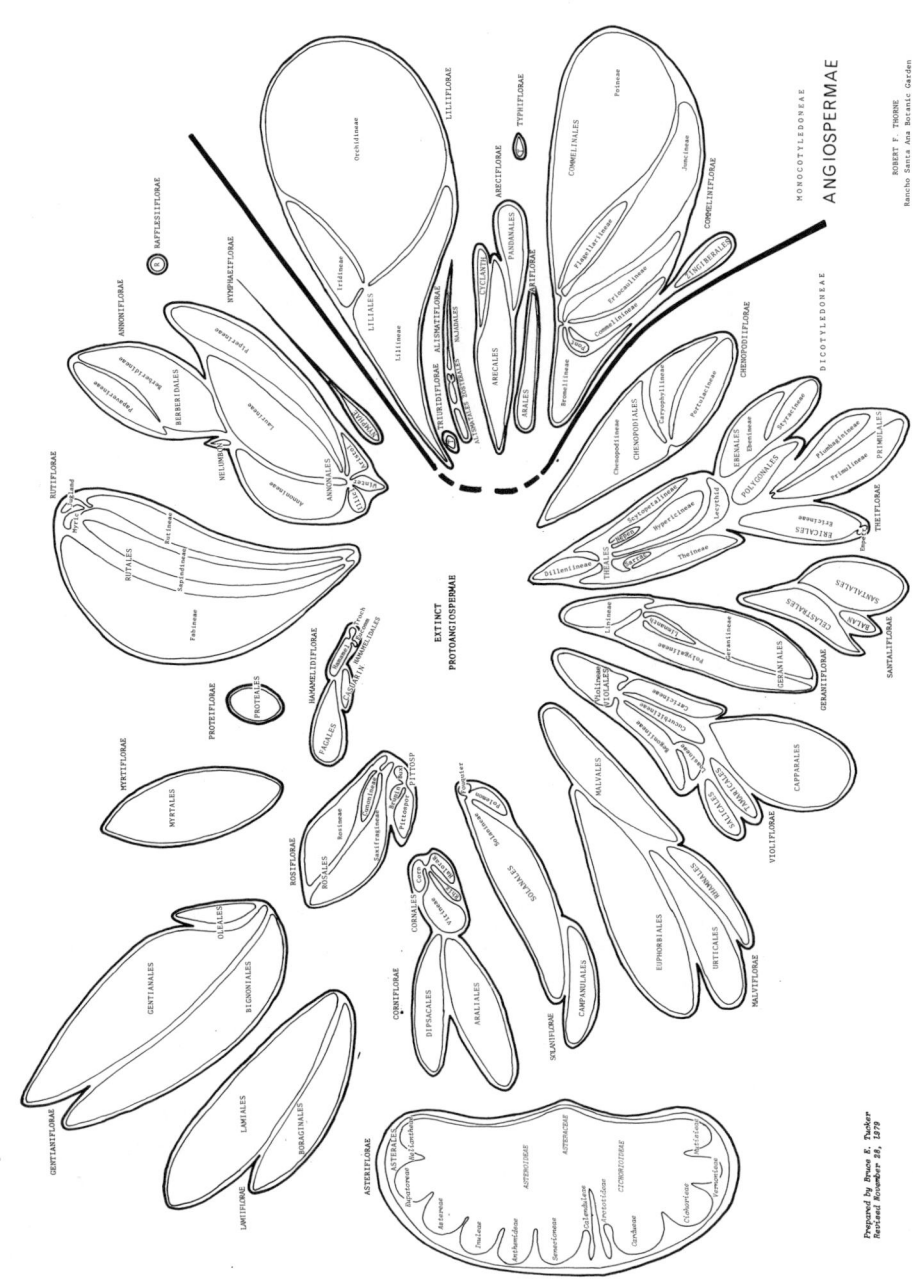

Figure 1. Diagrammatic representation of the phylogenetic relationships of the angiosperms (see text for a detailed explanation).

With these caveats taken care of, I should like to turn now to my new phyletic diagram (Fig. 1). It is an attempt to depict my classification of the angiosperms as they exist at this point in time. The balloons, or bubbles if you prefer, indicate by their size the approximate number of extant species. The lobes indicate orders and the constrictions separate suborders. Only the huge Asteriflorae-Asterales-Asteraceae are divided further into the two subfamilies and most commonly recognized tribes, my concession to the syngenesiologists that abound throughout the botanical world. Some tiny taxa are greatly enlarged here relative to their true size in order for us to see them, especially the Nelumbonales, Triuridiflorae, Typhiflorae, and Rafflesiiflorae. The less specialized taxa are closest to the empty center, representing the long extinct and vast group of protoangiosperms that conquered the terrestrial world in Cretaceous times. The most specialized taxa are found toward the periphery of the diagram, the longer balloons indicating considerable evolutionary depth. Although my diagram has now gone through many permutations, I am still far from satisfied with it. The Annoniflorae and Nymphaeiflorae should jut farther into the center; the Hamamelidiflorae should be reoriented to stretch the Fagales farther to the periphery; the Caryophyllineae, Limnanthineae, and Caricineae should be scrapped; the Paeoniales should be added to the Annoniflorae near the Berberidales; and the Salicales and Tamaricales should be dropped to subordinal status in the Violales. The flow of new phyletic information is so voluminous and my system so elastic that my artists cannot keep up with my phyletic thinking. This diagram is by no means intended to represent a phylogenetic tree. I do not believe such a tree is a realistic possibility at this time, if ever. The most realistic representation attempting to depict angiosperm lineages would appear more like a phyletic shrub, or better still, a phyletic hedge.

ANNONIFLORAE-NYMPHAEIFLORAE

Using this phyletic diagram, I should like to discuss some of my recent realignments, emphasizing those supported by phytochemical evidence. Going first to the Annoniflorae, which contains the greatest assemblage of taxa abounding in ancestral characteristics, we have learned already in this symposium that this group, essentially the Polycarpicae of earlier days, is rather well defined generally by common possession of benzyl-isoquinoline alkaloids, sesquiterpene lactones, polyacetylenes, flavonols, and tyrosine-derived cyanogens and by the general absence of ellagic acid, myricetin, and iridoids (Dahlgren 1981, Saupe 1981, and Seigler 1981, in this symposium volume; also Bate-Smith 1968, 1974; Dahlgren 1977a; Gornall, et al. 1979; Hegnauer 1971, 1977; Kubitzki 1969; Rezende, et al. 1975; Seigler 1977; and Valen 1978). There can be little argument about the common ancestry of the Annonales, Nelumbonales, and Berberidales.

Nelumbonales-Nymphaeales

There may, however, be some surprise over the removal of *Nelumbo* Adans. from the Nymphaeales and the treatment of the Nymphaeales in a separate superorder Nymphaeiflorae. Despite the similarities in habitat and growth habit of *Nelumbo* and the Nymphaeales which seemed to indicate their close relationship, I was long troubled by the differences in stem anatomy (*Nelumbo* has primitive vessel elements, the Nymphaeales no vessels or rather specialized ones in *Cabomba* Aubl. (Inamdar and Aleykutty 1979)); pollen grains (*Nelumbo* has tricolpate grains; the Nymphaeales monosulcate or monosulcate-derived graints); seedlings, etc. However, the serological and other phytochemical differences have proved decisive. My former colleague Jean-Pierre Simon (1970, 1971) in his serological studies of the water-lilies stressed the slight serological relationships between *Nelumbo* and Nymphaeaceae; whereas, he found partial identity reactions of *Nelumbo* with members of the Annonales and Berberidales, including the Papaveraceae. The Nymphaeales lack benzylisoquinoline alkaloids but have sesquiterpene alkaloids, ellagitannins, and myricetin (Bate-Smith 1974; Gornall, et al. 1979; Hegnauer 1971; Seigler 1977). Like the Annoniflorae, Rutiflorae, Hamamelidiflorae, and monocotyledonous superorders, the Nymphaeiflorae retain many characteristics, both morphological and chemical, from their distant and long-extinct protoangiospermous ancestors, but they certainly are not closely related to nor derivable from other extant superorders.

Berberidales

The Berberidineae include many largely herbaceous Asiatic relicts. As much as I dislike to accept a multiplicity of monogeneric, especially monospecific, families, the imperative to recognize monophyletic taxa and to attain equivalence of taxonomic treatment in my classification has led me to accept several more small families in this suborder. *Nandina* Thunb. as the Nandinaceae (Kumazawa 1938; Shen 1954); *Glaucidium* Sieb. & Succ. as the Glaucidiaceae (Tamura 1972); and *Circaeaster* Maxim. and *Kingdonia* Balfour f. & W. W. Smith as the Circaeasteraceae (Junell 1931; Foster 1959, 1961a, 1961b, 1963; Foster and Arnott 1960). All possess classificatory gaps between them and the presumably related Berberidaceae and Ranunculaceae as great as those between widely accepted, larger and less relict dicotyledonous families.

Paeoniales

The removal of *Paeonia* L. from the Ranunculaceae and acceptance of the genus as an independent family is supported by the lack of serological affinity with the Ranunculaceae (Hammond 1955; U. Jensen 1968; Fairbrothers, et al. 1976) and by wood anatomy and other comparative morphology (Keefe and Mosely 1978). The prevalence of centrifugal stamen initiation in various investigated species of *Paeonia*, the large follicular fruit, and other features led Corner (1946) and others, including myself, to treat the Paeoniaceae near the Dilleniaceae. However, like Corner (1976), I have been troubled by the marked differences between *Paeonia* and the Dilleniaceae in seeds, pollen grains, leaves, habit, distribution, and other features. The report of centripetal development of stamens in *Paeonia japonica* (Makino) Miyabe & Takeda by Sawada (1971), the berberid-ranunculid-papaverad dissection of the leaves, and other probably basic resemblances to members of the Berberidales have caused me tentatively to transfer the Paeoniaceae back to the Annoniflorae, but as the digeneric order Paeoniales adjacent to the Berberidales. *Glaucidium palmatum* Sieb. & Zucc., also a rhizomatous perennial with centrifugal initiation and maturation of stamens, rather unspecialized wood anatomy, and follicular fruit, seems to deserve coordinate family status in the order as the Glaucidiaceae (Tamura 1972). The stem anatomy of *Paeonia* is much less specialized than that of any member of the Berberidales. I believe that *Paeonia* and *Glaucidium* are relicts of a group having common ancestry with the Berberidales, but retaining more archaic features of stem anatomy, a much lower chromosome number (n=5 in *Paeonia*), primarily centrifugal stamen initiation, and massive follicles, among other distinctive features.

THEIFLORAE

Dilleniineae-Theineae-Sarraceniineae

As discussed above, I have removed the Paeoniaceae from my Dilleniineae. Although Keefe and Mosely (1978) favor retaining the Paeoniaceae near the Dilleniaceae, they, like Schmid (1979a), stress the close relationship of the Dilleniales with the Theales, which I indicate by placing the Dillenineae as the least specialized suborder of my Theales. Schmid agrees with this treatment and also prefers placing the Actinidiaceae in the Theineae. The presence of stamen fascicle traces definitely excludes the Actinidiaceae from the related Ericales, where Dahlgren (1975) places the family because of its production of iridoids. A later paper by Schmid (1979b) also disposes of any relationships of the Actinidiaceae with the nyssaceous *Davidia* Baill., but does support the relationship of the Actinidiaceae with the Paracryphiaceae.

Excellent studies by Baas (1974, 1975), Carpenter and Dickison (1976), and Dickison and Baas (1977) have convinced me that *Phelline* Labill., *Oncotheca* Baill., *Paracryphia* Baker f., and *Sphenostemon* Baill., along with the Icacinaceae and *Peripterygium* Hassk. (Cardiopteridaceae), must be treated as separate families near the Theaceae and Aquifoliaceae in the Theineae. Likewise, the monographic work of Nooteboom (1975), added to Corner's comments (1976), has tipped the phyletic scales in favor of removing the Symplocaceae from the Ebenales to the Theineae. Studies of the Sarraceniaceae by DeBuhr (1975, 1977) place the Sarraceniaceae as the Sarraceniineae rather definitively next to the Theineae.

Scytopetalineae-Nepenthineae

The Old World pitcher-plants of the Nepenthaceae seem to have no close ties to the New World pitcher-plants, but instead have, as the Nepenthineae, evolved separately and are placed in the Theales near the Scytopetalineae (Thorne 1977). In regard to the putatively related Dioncophyllaceae and Ancistrocladaceae (which I have transferred from the Geraniales to the Scytopetalineae (Thorne 1977)), Hegnauer (pers. comm.) has called my attention to the additional evidence that members of both families possess both acetogenic isoquinoline alkaloids of characteristic structure (ancistrocladine-type alkaloids) and biogenetically related naphthoquinones.

My concept of the Scytopetalineae has been changed somewhat by the removal from it of the Dipterocarpaceae and Sarcolaenaceae to the Malvales. The recognition of the dipterocarps and sarcolaenads as malvalean is necessitated by the detailed study of the tiliaceous *Pakaraimaea* Maguire by Maguire (1977), Maguire and Ashton (1977), Zeeuw (1977), Giannasi and Niklas (1977), and by Kostermans (1978); the work on dipterocarp pollen exines by Maury, et al. (1975); and the investigations of seed structure by Corner (1976). Another change in the Scytopetalineae is the transfer to the suborder of the Medusagynaceae, a rare monotypic relict of the Seychelles. The strange umbrella-shaped septicidal capsule of *Medusagyne seychellarum* Baker had previously influenced my treatment of the family as probably related to the Eucryphiaceae of the Rosales. However, examination of herbarium material of *Medusagyne* Baker has convinced me that the plant is indeed thealean and that the similarity of the fruit to that of *Eucryphia* Cav. is a convergence.

Lecythidineae

The position of the Lecythidineae in the Theales has been strengthened recently by evidence from the seeds (Corner 1976) and leaves (Hickey and Wolfe 1976), and by their exclusion by most workers concerned with the Myrtales, largely because of their lack of internal phloem, vestured pitting and the centrifugal initiation of the usually numerous stamens. The family also has received much attention from Kowal, et al. (1977), Mori, et al. (1978), Prance (1976), Prance and Mori (1977, 1978), and Whitmore (1974).

Ericales

Few would doubt the close relationships of the Ericales to the Theales although there is still much quibbling as to the circumscription of the order, which I prefer to limit to the Ericaceae, Epacridaceae, and Empetraceae. Watson (1964, 1965), Stevens (1971) and Wallace (1975, 1976, 1977) have contributed greatly to a better definition of the Ericaceae to include the Pyroloideae, Monotropoideae, and Vaccinioideae. I prefer to retain the Wittsteinioideae in the Epacridaceae. Harborne and Williams (1973) made an informative survey of the flavonoids and simple phenols in the leaves of the Ericaceae and their suggested relatives.

Primulales

The Primulales have been somewhat more controversial, for some of us, as Hutchinson (1973), prefer to include in the order the Plumbaginaceae, preferably as the separate suborder Plumbaginineae (Thorne 1976, 1977). Evidence from the seeds (Corner 1976) and flavonoids (Fairbrothers, et al. 1976; Gornall, et al. 1979) seems to support this treatment. Anthocyanins are reported only from the Primulaceae and Plumbaginaceae, aside from the quite unrelated Gesneriaceae (Fairbrothers, et al. 1976). The S-type sieve-element plastids of both the Plumbaginaceae and Polygonaceae (Behnke 1976d) added to the absence of betalains and centrospermous seed features definitely remove both families from the Chenopodiales (Centrospermae). Harborne (1967a, 1969a) lends phytochemical support by his listing of the related europetin in *Plumbago L.* and hirsutidin in *Primula L.*; plumbagin in all examined Plumbaginineae and in the Ericaceae; and flavonol 5-methyl ethers in three genera of Plumbaginaceae, eight genera of Ericaceae, and *Tetracera L.* of the Dilleniaceae, as well as in three apparently unrelated families (Eucryphiaceae, Juglandaceae, and Lauraceae).

Polygonales

My transfer of the Polygonales to a position near the Primulales is tentative. The presence of ellagitannins in both Polygonaceae and Plumbaginaceae (Bate-Smith 1974), of starch in the pollen grains (Baker and Baker 1979), and both S-type sieve-element plastids mentioned above for both families (Behnke 1976d), and the absence of betalains and perisperm, are evidence backing this placement. The tricolpate to polyporate, trinucleate pollen grains; syncarpous, unilocular pistil with only one basally-attached, bitegmic, crassinucellate ovule; fruit usually a small nut; and seed with starchy endosperm may also indicate common ancestry or may be convergences. The two families have many troublesome differences, and if related, they must have diverged greatly from their putative common ancestors. Dahlgren (1975) agrees with this removal of the Primulaceae, Plumbaginaceae, and Polygonaceae from the Chenopodiiflorae and their treatment as related orders.

CHENOPODIIFLORAE

The Chenopodiiflorae are now a very well-defined group that Young in this volume (1981) and I seem to agree are related, at least distantly, to the Theiflorae. The removal of numerous extraneous elements, as *Thelygonum* L. to the Rubiaceae (Wunderlich 1971), Gyrostemonaceae and Bataceae to the Sapindineae (Thorne 1977), and Polygonaceae, as discussed above probably to the vicinity of the Primulales, has left a solid core of closely related families so much studied (Behnke 1974, 1975a, 1975b, 1976a, b, c, d, e, 1977, 1978; Behnke, et al. 1974, 1975; DiFulvio 1975; Eckardt 1976; Ehrendorfer 1976a, b; Fairbrothers, et al. 1976; Gerdemann 1968; Gibson 1973, 1978; Gibson and Horak 1978; Hunziker, et al. 1974; Mabry 1974, 1976, 1977; Mabry and Behnke 1976; Mabry, et al. 1978; Richardson 1978; Skvarla and Nowicke 1976; Seigler 1977; and Yoong et al. 1975) as to need little discussion here. I am still, however, rather concerned about the internal classification of the Chenopodiales. I have re-examined the Molluginoideae, and despite their production of anthocyanins rather than betacyanins, I am unable to separate morphologically the subfamily from the betalain-producing aizoads. Thus, I am returning the molluginoids to the Aizoaceae, and treating the Caryophyllaceae, also anthocyanin-producing, immediately adjacent to the Molluginoideae. Thus, I have submerged the Caryophyllineae, as well as the Phytolaccineae, in the Chenopodiineae. The apocarpous *Gisekia* L. seem to me best removed from the Molluginoideae and placed as the Gisekioideae in the Phytolaccaceae. Mabry in Fairbrothers, et al. (1976) has reported betalains in *Gisekia*. The genus needs further study as do the true molluginoids. *Halophytum* Spegazz. has been studied recently by Di Fulvio (1975), Gibson (1978), and Hunziker, et al. (1974). This monospecific genus seems best placed as the Halophytaceae near the Aizoaceae. The digeneric Hectorelloideae, studied by Behnke (1975c, 1977), Mabry, et al. (1978), and Yoong, et al. (1975), are certainly centrospermous and seem reasonably at home in the Portulacaceae but are very specialized cushion-plants.

GERANIIFLORAE

The Geraniiflorae, whose Linineae appear to have retained many common features with the Theiflorae from presumed common ancestors, remain much as I have treated them earlier (Thorne 1976) except for the removal of the Ancistrocladaceae discussed above and the recognition in the Linineae of the Houmirioideae and Ctenolophonoideae as the distinct families Houmiriaceae and Ctenolophonaceae, considered somewhat less specialized than the Linaceae. Corner's seed investigations (1976) show the Zygophyllaceae to be out of place in the Sapindales, where some taxonomists would place them. Corner suggests placement with the Malpighiales (my Polygalineae of the Geraniales); but I think the family fits better in the Linineae. Seed structure (Corner 1976) also eliminates *Balanites* Delile from the Zygophyllaceae and favors treatment with the Simaroubaceae, where I place them as the Balanitoideae. In the Geraniineae Lefor (1975) and Goldblatt (1979) have presented sufficient evidence to convince me that the Ledocarpaceae and Vivianiaceae of South America are distinct enough from the related Geraniaceae to merit separate family status.

SANTALIFLORAE

The relationship of the Santaliflorae via the Celastraceae to the Geraniiflorae, most likely through common ancestry with the Linineae, is quite problematic, but I cannot suggest at this time a better position for them. I am hoping that our student, Barry Prigge, in his monographic study of *Mortonia* A. Gray of the Celastraceae, can ultimately place the family and its relatives more definitively. I have removed the Icacinaceae and probably related Cardiopteridaceae from the Santaliflorae to the Theineae as discussed above, thus leaving, I hope, the stripped-down Celastrales as a natural unit. As usually treated by most taxonomists, the Celastrales have been probably the most ridiculously unnatural assemblage among the dicotyledons, a veritable biological waste basket. Whether the Santalales and Balanophorales are truly related to the Celastrales and to each other is also a matter for much further study and discussion.

VIOLIFLORAE

The common occurrence of glucosinolates (Ettlinger, pers. comm.; Kjaer 1974; and Rodman 1981) in the Limnanthaceae and Tropaeolaceae of the Geraniales, Caricaceae of the Violales, and Brassicaceae, Capparaceae (including *Pentadiplandra* Baill. and *Tovaria* Ruiz et Pav.), Moringaceae, and Resedaceae of the Capparales may have some significance in indicating close relationships among these three orders. However, we must keep in mind that glucosinolates, isothiocyanate precursors, myrosinase, myrosin cells, or related compounds or structures, have also been found in the Sapindineae, *Drypetes* Vahl of the Euphorbiaceae, Salvadoraceae, and other presumably unrelated taxa. Myrosin cells have recently been found in the capparaceous genus *Koeberlinia* Zucc. by Gibson (1979), which because of the axile placentation is perhaps best treated as a separate subfamily Koeberlinioideae, much as *Tovaria* in the Tovarioideae. Also probably of considerable significance is the prevalence of centrifugal stamen initiation among the pluristaminate members of most of these orders as well as in the apparently related Theiflorae, Chenopodiiflorae, and Malviflorae (Corner 1946). The occasional centrifugal initiation of stamens found in pluristaminate taxa which normally have centripetal initiation or centripetal development in groups normally with centrifugal development (Sawada 1971; Eyde 1976; Uhl and Moore 1977) helps to remind us that few, if any, features are monophyletic.

Violales

The Cistaceae, which I formerly treated in the Violales, and which lent their name via Lindley (1833) to the order, suborder, and superorder, must be transferred to the Malvales near the closely related Bixaceae and Cochlospermaceae. Floral morphology, stem anatomy, and chemistry all combine to demand this realignment. Though closely related, the Bixaceae and Cochlospermaceae apparently are sufficiently distinct to warrant family status (Dathan and Singh 1972; Keating 1969, 1970, 1972, 1974). The ordinal name Violales, also used by Lindley (1833) is now appropriate for this order and also provides sub- and superordinal names, Violineae and Violiflorae.

The Salicaceae, despite the basic anemophily of *Populus* L. and secondary entomophily of *Salix* L., approach the Flacourtiaceae, especially the temperate Asiatic *Idesia polycarpa* Maxim. of the Idesideae, so closely that they can be separated from the latter family at most as a suborder, the Salicineae, placed near the related Tamaricineae between the Violineae and Cucurbitineae. I previously had treated both taxa as orders with close common origin with the Violales. Meeuse (1975) found a positive association between the Salicaceae and Flacourtiaceae in respect to both embryological and chemical characteristics; and Hickey and Wolfe (1975) emphasized the similarities of the leaves of the Salicaceae and *Idesia*, especially the salicoid teeth, actinodromous venation, and stipules.

The Loasaceae are a most troublesome family, whose position is much debated. I continue to retain it as a very specialized monofamilial suborder in the Violales, but without strong convictions for that placement. Bisexual, radially-symmetrical, mostly polypetalous flowers; girdling vascular bundles in the flowers; centrifugal stamen development; and parietal placentation seem to argue for retention in the Violales and against the recently popular placement as the Loasanae (Dahlgren 1975, 1977a, b) among the Sympetalae, though the presence of iridoid glycosides (Kooiman 1974) and the unitegmic, tenuinucellate ovules with endosperm cellular ab initio, with terminal haustoria (Dahlgren 1977a, 1977b) do argue for great embryological specialization and distant, if common, ancestry with the other Violales. The Loasaceae certainly require much further intensive study to get definitive evidence for their relationships.

MALVIFLORAE

The Malviflorae, much changed by the removal of the Solaniflorae and addition of the Dipterocarpaceae, Sarcolaenaceae, and Cistaceae, and most recently by the addition of the Plagiopteraceae (Baas, et al. 1979), still remain a large group of rather closely related orders.

Urticales

Berg (1973, 1977) agrees that the Urticales are related to the Malvales and that the Moroideae, Cecropioideae, and Urticoideae are three very closely related groups. Recently (Berg 1978), he has elevated the Cecropioideae to family rank as the Cecropiaceae, certainly a more popular stance than retaining the three subfamilies in the Urticaceae. Yet, popular or not, I think the latter treatment is more realistic in view of the lack of familial-sized gaps among these closely related taxa.

Rhamnales

My treatment of the Rhamnaceae and Elaeagnaceae as the

sole members of the Rhamnales of the Malviflorae hardly has been very popular either. Nonetheless, the evidence for this treatment is becoming rather overwhelming. Corner (1976) unites the Rhamnaceae and Elaeagnaceae as exotestal in their seed-coats because of the tendency of the palisade-cells to have a stellate lumen, if not a more or less stellate facet. On the other hand, seed structure separates the Vitaceae because of their very primitive features, which I consider cornalean. Apparently, some of the parasitic fungi, like *Puccinia coronata* Cda. and *Plasmodiophora elaeagni* Schroet., the latter found in root nodules containing nitrogen-fixing endophytes, agree with me, with Hutchinson (1973) and with Corner (1976) for they parasitize members of the two families (Thorne 1979). The relationship of the Rhamnales and Euphorbiales is indicated by similar alkaloids found in the euphorbiaceous *Hymenocardia* Wall. ex Lindl., in the Pandaceae, and in the Rhamnaceae, but also in the Celastraceae (Seigler 1977).

Euphorbiales

I regret that I lapsed briefly from phyletic grace in my last published synopsis (Thorne 1976) by treating the Thymelaeaceae in the Myrtales, as some of my phyletic colleagues still do. However, I soon returned the Thymelaeaceae to the Euphorbiales, where diterpenes join morphology in assuring us that they belong there. The tigliane, daphnane, and ingenane diterpenes are known only from the Euphorbiaceae and Thymelaeaceae, with the daphnane diterpenes common to both families (Evans and Soper 1978; Ourisson 1974).

Also returned to the Euphorbiales, but without the quite unrelated Buxaceae, is *Simmondsia chinensis* (Link) C. K. Schneid. In his serological investigations of *Simmondsia* Nutt. my colleague Ron Scogin reported to me (pers. com.) that he had obtained weak reactions between the rabbit serum and two euphorbiaceous genera, *Ricinus* L. and *Jatropha* L. That phytochemical clue encouraged me to compare *Simmondsia* with the Euphorbiaceae, especially with the genus *Colliguaja* Molina of temperate South America. Practically all the anatomical and morphological characteristics of *Simmondsia* can be found among the Euphorbiaceae, but never in the same combination. I am now rather convinced that *S. chinensis*, the jojoba, should be treated as the family Simmondsiaceae in close proximity to the Euphorbiaceae.

RUTIFLORAE

The position and content of my Rutiflorae may surprise some botanists. In fact, the removal of the Fabineae from the Rosiflorae to the Rutiflorae may even shock the more conservative members of our taxonomic community. However, such tradition-bound individuals need to be jarred out of their phyletic apathy occasionally. I am reminded of a statement made by Watson (1964) that particularly delighted my iconoclastic mind, "Too many modern botanists seem to accept traditional systems uncritically and, unwittingly, to base their conclusions on nineteenth-century taxonomic philosophy."

Rutineae

No taxonomic group is more closely related chemically to the Annoniflorae than the Rutaceae. Not only are most members of the family characterized by essential oils in lysigenous cavities, similar to those in the Annonales, but some of the least specialized, ancient genera like *Zanthoxylum* L., *Phellodendron* Rupr., and *Toddalia* Juss. have benzylisoquinoline alkaloids like those in the Annoniflorae (Seigler 1977; Waterman 1975). The limoids, citrus bitter principles, are rather evenly distributed throughout the closely related Rutaceae and Meliaceae, with the closely related simaroubolides characteristic of the equally near-related and often apocarpous Simaroubaceae (Dreyer 1966). Essential oils are found in secretory cells in the Cneoraceae, Simaroubaceae, and Meliaceae and in schizogenous canals in the Simaroubaceae, Burseraceae, and Anacardiaceae (Hegnauer 1978a). The Rutineae are thus rather closely knit biochemically. I have returned *Rhabdodendron* Gilg & Pilger as the Rhabdodendroideae to the Rutaceae, where I should have left them. As soon as I examined material of the genus, supplied kindly by Ghillean T. Prance, I observed at once the characteristic rutaceous lysigenous secretory cavities in the leaves and other organs, which appear against the light as pellucid dots (see Puff and Weber 1976).

Juglandineae, Myricineae, Leitneriaceeae

I have discussed elsewhere (Thorne 1974a) the Juglandineae, Myricineae, and *Leitneria* Chapm., now placed as the Leitneriaceae in my Rutineae near the Anacardiaceae, but I might remind the reader here of some of the morphological and chemical similarities of these taxa to the Rutineae. Juglandinean leaves, like those of many Rutineae, are basically alternate, exstipulate, imparipinnately-compound, glandular, aromatic, resinous, and supplied with anomocytic stomata and resin-secreting peltate glands. The woody stems have secretory cells and clustered crystals and the bark is rich in tannins. The aromatic but simple-leaved Myricineae are biochemically similar (Dahlgren 1975), and the less known Leitneriaceae have secretory canals with yellow resinous contents in stem pith margins and in leaf midveins and also have bark rich in tannins. I have recently examined specimens of *Canacomyrica* Guill. collected on Mt. Panié, New Caledonia, by Sherwin Carlquist; and I agree with him that the genus is certainly myricaceous. All these taxa are ancient and present evidence of bisexual ancestry. Young (1976, 1981) has used flavonoid chemistry along with stem anatomy and other morphology to reduce the formerly recognized amentiferous order Julianiales to subtribal status, Julianiinae, in the anacardiaceous Rhoeae, the sumac tribe. Serological work by Petersen and Fairbrothers (1979) agrees with him. The anacardiaceous affinities of these amentiferous groups can no longer be lightly denied. It is worth repeated here that the basic chromosome number of the Juglandaceae, Myricaceae, and *Leitneria*, like that of many Anacardiaceae, is x = 16 (Bolkhovskikh, et al. 1969; Raven 1976). Even the subordinal rank of the Juglandineae and Myricineae may exaggerate the differences between these taxa and the Rutineae.

Sapindineae

Another important group in the Rutales is the Sapindineae, especially the basic, very significant family Sapindaceae surrounded by numerous small, relict, satellite families like the Akaniaceae, Bretschneideraceae, Aceraceae, and Hippocastanaceae. When the Gyrostemonaceae and Bataceae were rightly cast out of the Chenopodiales (Centrospermae), the presence of glucosinolates in the gyrostemonaceous *Codonocarpus* A. Cunn. ex Hook. and the enzyme catalyzing hydrolysis of indolyl methy-glucosinolate in *Batis* P. Br. (Kjaer 1974) induced several botanists to force these two related families into or near the heavily glucosinolate Capparales. Those botanists overlooked the fact that myrosin cells, associated with glucosinolates, were reported long ago in *Bretschneidera* Hemsl. and suspected in *Akania* Hook. f., the turnipwood of the Australians, both closely related to the Sapindaceae. As reported elsewhere (Thorne 1977), I consider that the basically Australian Gyrostemonaceae, Bataceae, and sapindaceous Dodonaeoideae, *Stylobasium* Desf., and *Emblingia* F. Muell. form an anemophilous complex with immediate common protosapindaceous ancestry. Carlquist (1978a) has recently discussed the wood anatomy and relationships of these taxa.

Fabineae

My Fabineae, with Connaraceae, Surianaceae, and Fabaceae, need some comment. The legumes have traditionally been placed in the Rosales near the Rosaceae, presumably because both families have basically apocarpous gynoecia, somewhat obscured in the legumes by general reduction of the pluricarpellate condition to the usual single carpel, ripening into a follicle or more commonly into a legume. Yet the Fabaceae seem to have little to relate them to the basically simple-leaved Rosaceae. According to Seigler (1977), the legumes have no secondary plant compounds that establish a close relationships with the Rosaceae. Saupe (1981) in this volume reports that leucine-derived cyanogens are restricted primarily to the Fabaceae and Sapindaceae, but also occur in at least one genus of the Rosaceae. Unlike most of the Rosales, the Fabaceae and closely related Connaraceae resemble the Rutineae and Sapindineae in having basically alternate, pinnately compound leaves and often arillate seeds. The seed structure of the Connaraceae, according to Corner (1976) is essentially like that of the Meliaceae and Sapindaceae.

PROTEIFLORAE

The Proteales and their single family Proteaceae have been returned in my present synopsis to their own superorder Proteiflorae, which I had originally suggested for them (Thorne 1968) to emphasize their isolation. With the removal of the Fabineae from the Rosiflorae, the Proteales also had to be removed, for I believe they have their closest living relatives among the Fabaceae. Like the legumes the proteaceous gynoecium consists of a single carpel, often stipitate, maturing into a fruit that is often a dry, dehiscent follicle. The stomata are, like those of the Fabaceae and most of the Connaraceae, paracytic, not anomocytic like rosaceous stomata. Also, though the mature leaves are often simple or dissected, the immature leaves of the more primitive, rainforest protead species are often imparipinnately compound. Dahlgren (1975) has considered the chemical contents of the Proteaceae as "somewhat reminiscent of those" in the Fabaceae. Johnson and Briggs (1975) consider the Proteiflorae to be an isolated group.

ROSIFLORAE

The Rosiflorae, with the removal of the Fabineae and Proteiflorae, are now a much smaller but still significant group containing often small, apparently relict taxa with an abundance of primitive features.

Rosales

At the Rancho Santa Ana Botanic Garden, we have recently examined the western American Crossosomataceae (DeBuhr 1978; Thorne 1977; Thorne and Scogin 1978), redefining the family after the recent addition of two more genera, *Apacheria* Mason and *Forsellesia* Greene. From his study of the biochemistry of the group, Scogin (Thorne and Scogin 1978) agrees with its placement in the Rosales, rather than in the Dilleniales as treated by Takhtajan (1969) and Cronquist (1968). It was apparently misplaced in the Dilleniales in the mistaken belief that the stamens are initiated centrifugally. *Crossosoma californicum* Nutt., the wild-apple of Santa Catalina Island, is held in rather high esteem by Southern California Botanists, furnishing the logo for their society and the name *Crossosoma*, for their journal.

Pittosporales

The Pittosporales recently have received much attention from my colleague Sherwin Carlquist (1975a, 1976a, 1976b, 1976c, 1977b, 1977c, 1977d, 1978b, 1978c) as a result of his own field investigations of the group in Australasia, Malaysia, and South Africa. I have found his studies, particularly of the wood anatomy, of the Pittosporineae and Brunineae to be most encouraging in my treatment of the group. In the appended synopsis, I have combined the Daphniphyllineae with the Buxineae, thus treating the Daphniphyllaceae, Balanopaceae, and Buxaceae under the older subordinal name Buxineae. I discussed this group at some length recently (Thorne 1977). The Pittosporales still need much further study to refine the relationships among the suborders and the position of the order anent the surely closely related Rosales, Hamamelidales, Cornales, and Araliales (Hegnauer 1969, 1978b). We do not yet know the significance of the presence of iridoids in such members of the order as *Daphniphyllum* Blume and *Roridula* Burm. f. ex L., though they have also been reported in *Eucommia* Oliv. and *Liquidambar* L. of the Hamamelidales, several saxifragaceous genera of the Rosales, and generally in the Cornales (including *Hippuris* L.) and Dipsacales (including the Calyceraceae) (Jensen, et al. 1975).

MYRTIFLORAE

The Myrtales have, like the Pittosporales, lately received much attention, especially from the anatomists (Baas 1979; Baas and Zweypfenning 1979; Beusekom-Osinga and Beusekom 1975; Bridgwater and Baas 1978; Briggs and Johnson 1979; Carlquist 1975b, 1977a; Carlquist and DeBuhr 1977; Graham 1975, 1978; Muller 1975; Schmid 1980; Skvarla, et al. 1975, 1976, 1978; Vliet 1974; and Vliet and Baas 1975). All this activity has refined considerably our circumscription of this rather natural order, with the elimination from it of such extraneous taxa as the Haloragaceae, Lecythidaceae, Rhizophoraceae, and Thymelaeaceae.

GENTIANIFLORAE

Some of the more startling realignments that I have made since my last published synopsis (Thorne 1976) include my splitting away of the Oleales from the Santaliflorae and of the Solaniflorae from the Malviflorae, with transfer of these taxa into closer proximity with the other sympetalous superorders. These radical shifts were largely occasioned by the accumulation of new phytochemical and other knowledge about these groups. Also, I had earlier over-reacted against the artificiality of the old Sympetalae.

Oleales

The Oleaceae, rich in iridoids derived from secologanin and in complex alkaloids (Jensen, et al. 1975), fit well near the Gentianales (including the Rubiaceae). In addition, they share with the Gentianales opposite leaves, determinate inflorescences, actinomorphic, tetracyclic, and sympetalous flowers, bicarpellate gynoecia, and few or solitary ovules in each of the two locules, the ovules unitegmic and tenuinucellate. The positioning of the Salvadoraceae in the Oleales is less certain, perhaps even speculative, but worthy of consideration by the experts.

Gentianales

The Gentianales need little comment, though there still remains much controversy as to the level at which various taxa are to be treated. Some would prefer to dismantle the diverse Loganiaceae rather completely though I think only the Buddlejoideae are distinct enough to warrant family status. In some respects the buddlejoids are as close to the Scrophulariaceae as to the Loganiaceae. *Retzia* Thunb. of South Africa may deserve separate family status. I remain unconvinced, however, that the several asclepiad subfamilies are distinct enough from the other apocynad subfamilies to warrant the recognition of two families. Similarly, the recently described *Saccifolium* Maguire & Pires, despite unusual saccate, alternate leaves, hardly seem to me distinct enough from other gentians to deserve more than subfamilial status, if that, though Maguire and Pires (1978) have described the genus as a new family Saccifoliaceae, which they admit is closely related to the Gentianaceae.

Lamiiflorae

I have now divided the Lamiiflorae into two orders, Boraginales and Lamiales. Phytochemically, many of the Lamiales appear somewhat more closely related to the Bignoniales and Gentianales because of their similar iridoids (Jensen, et al. 1975) than to the non-iridoid-producing Boraginales. However, not all the Lamiales produce iridoids. Also the Lamiales and Boraginales do share basically cymose inflorescences and a mostly bicarpellate, superior ovary that commonly is tetraloculate by false partitions and deeply four-lobed, and develops into a drupe with four one-seeded pyrenes or a cluster of four one-seeded nutlets. They also share strong tendencies toward an herbaceous stem bearing alternate or opposite, exstipulate, simple leaves; a gynobasic style; reduction in ovule number to one, and basally attached in each chamber of a deeply-divided ovary; and a seed with large embryo and no endosperm.

SOLANIFLORAE

The Solaniflorae, with the Solanales and Campanulales, mostly lack iridoids (except for the Fouquieriaceae and Goodeniaceae (Jensen, et al. 1975)) but are in general rich in caffeic acid or its derivatives. They are usually woody plants or herbs with alternate, exstipulate, mostly simple leaves and bisexual, pentamerous, showy, sympetalous flowers usually with five or ten epipetalous stamens and five- to two-carpellate gynoecia with many to one anatropous ovules on axile placentae. The Solanineae, often thought to be rather closely related to the Scrophulariaceae differ from them in their possession of internal phloem as well as marked differences in chemistry and floral anatomy. I have discussed the Fouquieriaceae elsewhere (Thorne 1977), and my colleague Ron Scogin has published on the floral anthocyanins (1977) and the leaf phenolics (1978) of the family. He found that the weight of phytochemical evidence favors affiliation with the Solanales. However, on the basis of the presence of iridoids, Dahlgren interprets differently the affinities of the Fouquieriaceae (Dahlgren, et al. 1976).

CORNIFLORAE

Essentially, the Corniflorae remain the same as in my last synopsis (Thorne 1976) except for the segregation of several genera from the previously heterogeneous Cornaceae and the separation of the Cornales (sensu stricto) and Araliales. The Cornaceae in the new narrower sense consist of *Mastixia* Blume (Mastixioideae), *Curtisia* Ait. (Curtisioideae), *Afrocrania* (Harms) Hutch. and *Cornus* L. (sensu lato) (Cornoideae), all with rather similar pollen grains (Ferguson 1977) and relatively similar chromosome numbers (2n = 22 or 26) (Goldblatt 1978). *Aucuba* Thunb., with n = 8 and very distinct intectate pollen grains (Ferguson 1977) and with closer chemical similarities, as aucubin, with *Garrya* Dougl. ex Lindl. (Hegnauer 1966), is best treated as the monogeneric family Aucubaceae in the Cornineae near the Alangiaceae and Garryaceae. *Helwingia* Willd. (n = 19) (Goldblatt 1978) and *Toricellia* DC. (n = 12) (Kurosawa 1977) have pollen grains unlike those of the Cornaceae (Ferguson 1977) and stem anatomy closer to the Araliaceae (Rodriguez 1971) and seem best treated as the Helwingiaceae and Toricelliaceae in the Araliales. The pollen grains, geography, and other features of *Kaliphora* Hook. f. and *Melanophylla* Baker suggest their inclusion with *Grevea* Baill. and *Montinia* Thunb. in the saxifragaceous, African-Madagascan Montinioideae (Ferguson 1977). The southern hemisphere *Griselinia* Forst. f., with pollen grains and vasculature somewhat similar to those of *Melanophylla* and multilacunar nodal condition (Neubauer 1978) probably are better placed with the Australasian *Corokia* A. Cunn. in the saxifragaceous Escallonioideae. All these relict genera should receive much more detailed study by specialists.

In their chemistry and serology, the Cornales and Dipsacales are closely allied; whereas, chemically, though not morphologically, they differ rather markedly from the Araliales (Hegnauer 1966; Bate-Smith, et al. 1975; Jensen, et al. 1975; Dahlgren 1977a; Gornall, et al. 1979). Their iridoids and ellagitannins, as well as their morphology, indicate relatively close common ancestry with the Rosales, Gentianales, and Lamiales, especially with the rosalean Saxifragaceae (sensu lato). The Araliales, on the other hand, show much chemical resemblance to the Asterales, Pittosporales, and Campanulales, especially in their sesequiterpene lactones and polyacetylenes, and their lack of iridoids, ellagitannins, and myricetin (Hegnauer 1969, 1978b; Dahlgren 1977a, 1977b; Gornall, et al. 1979). Thus, it seems desirable to express the chemical distinctness of the Araliineae from the other Corniflorae by elevating the group to ordinal rank as the Araliales.

MONOCOTYLEDONEAE

The appended synopsis contains few substantive changes among the Monocotyledoneae. Enough evidence has accumulated to cause me to recognize the Cymodoceeae of the Zannichelliaceae as a distinct family (Taylor 1909; Den Hartog 1970; Posluszny and Tomlinson 1977; Tomlinson and Posluszny 1976; Vijayaraghavan and Kumari 1974); Mapanioideae as a distinct subfamily in the Cyperaceae (Eiten 1976); Trithurieae as unrelated to the Centrolepidaceae (Hamann 1976, 1976) and listed as the Hydatellaceae in my taxa incertae sedis; and Strelitzioideae and Heliconioideae and Costoideae as families distinct from the Musaceae and Zingiberaceae (Tomlinson 1969). *Joinvillea* Gaudich., currently popular with the agrostologists as a presumed grass relative, is vegetatively quite distinct from *Flagellaria* L., but I remain unconvinced that it deserves greater status than subfamilial rank in the Flagellariaceae. Tomlinson and Smith (1970) have, however, elevated the genus to family rank as the

Joinvilleaceae.

TYPHIFLORAE

To indicate their rather isolated position, I have elevated the Typhaceae (including the Sparganioideae) to ordinal and superordinal rank, Typhales and Typhiflorae. My earlier belief that the Typhaceae were little more than anemophilous aroids was based, very tenuously it would seem, on similar inflorescences and a cross-inoculable rust parasite *Uromyces sparganii* Clint. & Peck of *Sparganium eurycarpum* Engelm. and *Acorus calamus* L. (Parmelee and Savile 1954). The bearing of the male flowers in spikes or heads above female flowers in similar spikes or heads is probably an expectable feature of the anemophilous syndrome. The parasitism of *Sparganium* L. and *Acorus* L., of similar habitat and growth habit, may well by an example of what my mycological colleague R. K. Benjamin (1967) has called fortuitous colonization (Thorne 1979). Lee and Fairbrothers (1972) found low serological correspondence between the Typhales and several Liliales and none with the Araceae or other families tested. They postulated that the Typhales might have originated from ancestral Liliales. However, I now agree with Cronquist (1968) and Dahlgren (1975) that a relationship with the Commeliniflorae is more likely considering the often common possession by both groups of growth habit as a glabrous perennial herb with starch-rich rhizomes; vessels in the stem and leaves as well as in the roots; stomata with two subsidiary cells; calcium oxalate often present as raphides or in other forms; the anemophilous syndrome, including one ovule in each of two or more locules of a compound ovary or a single ovule in a pseudomonomerous ovary; the ovule often apical, pendulous, anatropous, bitegmic, and crassinucellate; endosperm formation commonly helobial; and pollen grains occasionally in tetrads.

ACKNOWLEDGEMENTS

In the preparation of this paper, the phyletic diagram, and the appended synopsis, I am indebted to far more people for field work guidance, specimens, reprints, communications, and discussions than I can possibly list here. However, of the individuals who lately have been most helpful are Reino Alava of Turku, P. Baas and R. Hegnauer of Leiden, H.-D. Behnke of Heidelberg, Rolf Dahgren of Copenhangen, Christopher Davidson of Los Angeles, James Doyle, John Tucker and Grady Webster of Davis, Arthur Gibson of Tucson, Dana Griffin III of Gainesville,

Hugh Iltis of Madison, Siwert Nilsson of Stockholm, A. E. Orchard of Hobart, Ghillean Prance of New York, David A. Young of Urbana, and my present associates Sherwin Carlquist, Barry Prigge, and Ron Scogin at Claremont. Bruce Tucker, our former student now at the University of Washington, Seattle, devoted many hours and much effort toward depicting visually my phyletic ideas by constructing the phylogenetic diagram (Fig. 1). To all these botanists and the curators who have lent me specimens for study I am most grateful.

Also I wish to offer thanks to the students of my advanced course in angiosperm phylogeny who over the years at Claremont have patiently followed the twists and turns of my phyletic ramblings and have consistently offered encouragement through their enthusiams and deep interest as well as useful suggestions and information from their own researches into the relationships of various taxa.

APPENDIX

EXPLANATION OF THE SYNOPSIS OF THE CLASS ANGIOSPERMAE

The classification presented here is the latest version of my synopses published in Aliso (Thorne 1968) and in Evolutionary Biology, Vol. 9 (Thorne 1976). As in those synopses, I have carried the hierarchy of the class down to the subfamily level where appropriate, and in the Asteraceae down to the tribe because of the relatively large size of the asteraceous tribes. Subfamilies are listed because in the present inflationary climate many taxonomists fragment the natural family groupings to such an extent that my subfamilies are often the other phylogenist's families, or, in extreme cases, orders. My philosophy of classification is set down in some detail in the above-mentioned Evolutionary Biology paper (Thorne 1976) and will not be repeated here. Briefly, I use my extended hierarchy of subclasses, superorders, orders, suborders, families, and subfamilies to indicate in as realistic a way as possible both relationships and degrees of divergence from common ancestors without multiplying families and other taxa beyond reason. In attempting to be consistent and objective in the size of phyletic gaps or discontinuities that I accept to separate related groups in each major category, I have had to demote some traditional families while elevating others to a rank above their usual status. Within each family I have, for the most part, accepted the classification of those authorities that I regard as best informed and phyletically most realistic in their subdivision of the family.

The names chosen for the various taxa are according to the International Code of Botanical Nomenclature (Stafleu, et al. 1978), although the principle of priority is extended for all categories up to the class. Lindley's Nixus Plantarum (1833) is the point of departure for ordinal names, since Lindley there first applied consistently the ending -ales to generic roots. Synonyms or additional included or excluded taxa are given usually only where the names or circumscriptions of taxa deviate considerably from those in A. Engler's Syllabus der Pflanzenfamilien, Ed. 12, Vol. 2 (Melchior 1964).

For this synopsis, I decided to present to the reader some indication of the degree of confidence that I place in the alignment given, the circumscription accepted, or both, for each major category. Therefore, I have devised a simple "A," "B," "C" scale to indicate increasing confidence on my part, no matter how misplaced that confidence might be. "A," as used with Rafflesiiflorae and Polygonales, represents minimal confidence, little more than for those taxa listed in the taxa incertae sedis. "B," as used with Geraniiflorae, Triuridiflorae, Typhiflorae, Celastrales, Balanophorales, Trimeniaceae, Hydnoraceae, Paeoniaceae, etc., suggests limited confidence with some evidence that the alignment or circumscription is correct. "C" used rather generally throughout the synopsis implies that I am rather confident that the assembled data allow us to make a realistic placement and circumscription. ("D" is an empty category that might be used for such certainty as could result from divine revelation; I am still patiently awaiting same.)

In order to make the balloons in Figure 1 reflective of the number of species recognized in each major group, I had to get as reliable an estimate as possible of the number of generally accepted species. With those estimates in hand, I decided for the convenience of the reader to list after each higher taxon the number of genera and species (12/230 indicating 12 genera and 230 species) mostly drawn from Willis's Dictionary of the Flowering Plants and Ferns, 8th Ed. (Airy Shaw 1973), unless later or better information was available to me, as for the Annonaceae (Walker 1976), Arecaceae (Moore 1973), Asteraceae (Heywood, et al. 1977), Balanophoraceae (Hansen 1975), Bromeliaceae (Smith 1974), Garryaceae (Dahling 1978), Haloragaceae (Orchard 1975), Loasaceae (Poston and Thompson 1977), Meliaceae (Pennington and Styles 1975), Styracaceae (Spongberg 1976), Taccaceae (Drenth 1975), Tremandraceae (Thompson 1976), Velloziaceae (Smith and Ayensu 1974, 1976), etc. Usually species totals will not add up exactly for the larger categories because, to avoid spurious accuracy, I have rounded off the larger numbers of species to the nearest ten.

No linear sequence can approximate the probable branchings of a true phylogeny nor indicate the close common origins among the various superorders. Therefore, the superorders in the synopsis are separated by a gap to remind the reader that each superorder is a major line of development of the phyletic hedge. As each superorder terminates, the classification drops back down the evolutionary ladder to the start of the next major line of ascent. I hope that Figure 1, used with the synopsis, will indicate how I think the major groups are interrelated. At least a two-dimensional diagram is an improvement over a linear sequence. Nonetheless, space considerations still prevent a really adequate presentation of common origins and parallel development. Although Figure 1 is my most recent effort to present my ideas of angiosperm classification, I have already a number of improvements in mind. The best synopsis or diagram we can invent at this time is, of course, tentative and surely far short of the realistic presentation future generations of evolutionary botanists will attain. It is a target for specialists to shoot at, or a construction begging to be taken apart and reassembled by better informed builders.

The short list of taxa incertae sedis terminating the synopsis includes those taxa known to me that I am quite unable to place even with an informed guess. Probably the Rafflesiiflorae should be listed here as well. Surely, some revisionary taxonomists can suggest other candidates for the list of taxa probably misplaced in their own taxonomic groups. No pretense of completeness is intended here; the brevity of the list is merely an indication of my own ignorance of those undoubtedly numerous anomolous or misplaced taxa. I shall greatly appreciate suggested candidates for a future, more complete, list.

For the statistically inclined, I have totalled up the various taxa covered by the synopsis, as follows:

Statistical Summary

	Dicotyledoneae	Monocotyledoneae	Total Angiospermae
Species	173,530	51,540	225,070
Genera	9,634	2,605	12,239
Subfamilies	350	94	444
Families	290	50	340
(Subfamilies and undivided families)	548	122	670
Suborders	52	13	65
Orders	40	12	52
(Suborders and undivided orders)	79	21	99
Superorders	19	7	26

LITERATURE CITED

Adams, R. M., II, and G. W. Smith. 1977. An S.E.M. survey of the five carnivorous pitcher plant genera. Amer. J. Bot. 64:265-272.

Airy Shaw, H. K. (ed.). 1973. Willis's A Dictionary of the Flowering Plants and Ferns, 8th Ed. Cambridge: Cambridge University Press.

Baas, P. 1974. Stomatal types in Icacinaceae: additional observations of genera outside Malesia. Acta Bot. Neerl. 23: 193-200.

------. 1975. Vegetative anatomy and the affinities of Aquifoliaceae, *Sphenostemon, Phelline,* and *Oncotheca.* Blumea 22:311-407.

------. 1979. The anatomy of *Alzatea* Ruiz & Pav. Acta Bot. Neerl. 28:156-158.

------, R. Geesink, W. A. Van Heel and J. Muller. 1979. The affinities of *Plagiopteron suaveolens* Griff. (Plagiopteraceae). Grana 18:69-89.

------, and R. C. V. J. Zweypfenning. 1979. Wood anatomy of the Lythraceae. Acta Bot. Neerl. 28:117-155.

Baker, H. G. and I. Baker. 1979. Starch in angiosperm pollen grains and its evolutionary significance. Amer. J. Bot. 66:591-600.

Bate-Smith, E. C. 1962. The phenolic constituents of plants and their taxonomic significance. I. Dicotyledons. J. Linnean Soc. London Bot., 58:95-173.

------. 1968. The phenolic constituents of plants and their taxonomic significance. II. Monocotyledons. J. Linnean Soc. London Bot., 60:325-356.

------. 1974. "Systematic Distribution of Ellagitannins in Relation to the Phylogeny and Classification of the Angiosperms". In Chemistry in Botanical Classification, edited by G. Bendz and J. Santesson. Uppsala: Nobel Found.

------, I. K. Ferguson, K. Hutson, S. R. Jensen, B. J. Nielsen and T. Swain. 1975. Phytochemical interrelationships in the Cornaceae. Biochem. Syst. Ecol. 3:79-89.

Behnke, H.-D. 1974. Elektronenmikroskopische Untersuchungen an Siebröhren-Plastiden und ihre Aussage über die systematische Stellung von *Lophiocarpus*. Bot. Jahrb. Syst. 94:114-119.

------. 1975a. P-type sieve-element plastids: a correlative ultrastructural and ultrahistochemical study on the diversity and uniformity of a new reliable character in seed plant systematics. Protoplasma 83:91-101.

------. 1975b. Elektronenmikroskipische Untersuchungen zur Frage der verwandtschaftlichen Beziehungen zwischen *Theligonum* und Rubiaceae: Feinbau der Siebelement-Plastiden und Anmerkungen zur Struktur der Pollenexine. Plant. Syst. Evol. 123:317-326.

------. 1975c. *Hectorella caespitosa*: ultrastructural evidence against its inclusion into Caryophyllaceae. Plant Syst. Evol. 124:31-34.

------. 1976a. The bases of angiosperm phylogeny: ultrastructure. Ann. Missouri Bot. Gard. 62:647-663 (for 1975).

------. 1976b. Ultrastructure of sieve-element plastids in Caryophyllales (Centrospermae), evidence for the delimitation and classification of the order. Plant Syst. Evol. 126:31-54.

------. 1976c. Die Siebelement-Plastiden der Caryophyllaceae, eine weitere spezifische Form der P-Typ Plastiden bei Centrospermen. Bot. Jahrb. Syst. 95:327-333.

------. 1976d. A tabulated survey of some characters of systematic importance in centrospermous families. Plant Sys. Evol. 126:95-98.

------. 1976e. Sieve-element plastids of *Fouquieria*, *Frankenia* (Tamaricales), and *Rhabdodendron* (Rutaceae), taxa sometimes allied with Centrospermae (Caryophyllales). Taxon 25: 265-268.

------. 1977. Zur Skulptur der Pollen-Exine bei drei Centrospermen (*Gisekia*, *Limeum*, *Hectorella*), bei Gyrostemonaceen und Rhabdodendraceen. Plant Syst. Evol. 128:227-235.

------. 1978. Elektronenoptische Untersuchungen am Phloem sukkulenter Centrospermen (inkl. Didiereaceen). Bot. Jahrb. Syst. 99:341-352.

------, C. Chang, I. J. Eifert and T. J. Mabry. 1974. Betalains and P-type sieve-tube plastids in *Petiveria* and *Agdestis* (Phytolaccaceae). Taxon 23:541-542.

------, T. J. Mabry, I. J. Eifert and L. Pop. 1975. P-type sieve-element plastids and betalains in Portulacaceae (including *Ceraria*, *Portulacaria*, *Talinella*). Can. J. Bot. 53:2103-2109.

Benjamin, R. K. 1967. Laboulbeniales on semi-aquatic Hemiptera. *Laboulbenia*. Aliso 6(3):111-136.

Berg, C. C. 1973. Some remakrs on the classification and

differentiation of Moraceae. Meded. Bot. Mus. Herb. Rijksuniv. Utrecht 386:1-10.

------. 1977. Urticales, their differentiation and systematic position. Plant Syst. Evol., Suppl. 1:349-374.

------. 1978. Cecropiaceae a new family of the Urticales Taxon 27:39-44.

Beusekom-Osinga, R. J. van and C. F. van Beusekom. 1975. Delimitation and subdivision of the Crypteroniaceae (Myrtales). Blumea 22:255-266.

Bolkhovskikh, Z., V. Grif, T. Matvejeva and O. Zakharyeva. 1969. Chromosome Numbers of Flowering Plants, edited by A. Fedorov (Russian and English prefaces). Leningrad: Acad. Sci. U.S.S.R.

Bridgwater, S. D. and P. Baas. 1978. Wood anatomy of the Punicaceae. IAWA Bull. 1978/1:3-6.

Briggs, B. G. and L. A. S. Johnson. 1979. Evolution in the Myrtaceae--evidence from inflorescence structure. Proc. Linnean Soc. N.S.W. 102(4):157-256.

Carlquist, S. 1975a. Wood anatomy and relationships of the Geissolomataceae. Bull. Torrey Bot. Club 102:128-134.

------. 1975b. Wood anatomy of Onagraceae, with notes on alternative modes of photosynthate movement in dicotyledon woods. Ann. Miss. Bot. Gard. 62:386-424.

------. 1976a. Wood anatomy of *Myrothamnus flabellifolia* (Myrothamnaceae) and the problem of multiperforate perforation plates. J. Arnold Arbor. 57:119-126.

------. 1976b. Wood anatomy of Byblidaceae. Bot. Gaz. 137:35-38.

------. 1976c. Wood anatomy of Roridulaceae: ecological and phylogenetic implications. Amer. J. Bot. 63:1003-1008.

------. 1977a. Wood anatomy of Onagraceae: additional species and concepts. Ann. Miss. Bot. Gard. 64:627-637.

------. 1977b. A revision of Grubbiaceae. J. S. Afr. Bot.

43:115-128.

------. 1977c. Wood anatomy of Grubbiaceae. J. S. Afr. Bot. 43:129-144.

------. 1977d. Wood anatomy of Tremandraceae: phylogenetic and ecological implications. Amer. J. Bot. 64:704-713.

------. 1978a. Wood anatomy and relationships of Bataceae, Gyrostemonaceae, and Stylobasiaceae. Allertonia 1:297-330.

------. 1978b. Wood anatomy of Bruniaceae: correlations with ecology, phylogeny, and organography. Aliso 9:323-364.

------. 1978c. Vegetative anatomy and systematics of Grubbiaceae. Bot. Notiser. 131:117-126.

Carlquist, S. and L. DeBuhr. 1977. Wood anatomy of Penaeaceae (Myrtales): comparative, phylogenetic, and ecological implications. Bot. J. Linnean Soc. 75:211-227.

Carpenter, C. S. and W. C. Dickison. 1976. The morphology and relationships of *Oncotheca balansae*. Bot. Gaz. 137:141-153.

Corner, E. J. H. 1946. Centrigual stamens. J. Arnold Arbor. 27:423-437.

------. 1976. The Seeds of Dicotyledons. 2 vols. Cambridge, Eng.: Cambridge University Press.

Cronquist, A. 1968. The Evolution and Classification of Flowering Plants. Boston: Houghton Mifflin Co.

Dahlgren, R. 1975. A system of classification of the angiosperms to be used to demonstrate the distribution of characters. Bot. Notiser 128:119-147.

------. 1977a. A commentary on a diagrammatic presentation of the antiosperms in relation to the distribution of character states. Plant Syst. Evol., Suppl. 1:253-283.

------. 1977b. A note on the taxonomy of the "Sympetalae" and related groups. Publ. Cairo Univ. Herb. 7 & 8:83-102.

------, S. R. Jensen and B. J. Nielsen. 1976. Iridoid compounds in Fouquieriaceae and notes on its possible affinities.

Bot. Notiser 129:207-212.

------. 1981. "Relations Between Phytochemistry and Other Kinds of Characters in Angiosperms". In Phytochemistry and Angiosperm Phylogeny, edited by D. A. Young and D. S. Seigler (in press).

Dahling, G. V. 1978. Systematics and evolution of Garrya. Contr. Gray Herb. Harvard 209:1-104.

Dathan, A. S. R. and D. Singh. 1972. Development of embryo sac and seed of Bixa L. and Cochlospermum Kunth. J. Indian Bot. Soc. 51:254-266.

DeBuhr, L. E. 1975. Phylogenetic relationships of the Sarraceniaceae. Taxon 24:297-306.

------. 1977. Wood anatomy of the Sarraceniaceae; ecological and evolutionary implications. Plant Syst. Evol. 128:159-169.

------. 1978. Wood anatomy of Forsellesia (Glossopetalon) and Crossosoma (Crossosomataceae, Rosales). Aliso 9:179-184.

Den Hartog, C. 1970. The seagrasses of the world. Verhandelingen, Koninklijke Ned. Akad. Wetensh. Natuurkunde 59:1-275.

Dickison, W. C. and P. Baas. 1977. The morphology and relationships of Paracryphia (Paracryphiaceae). Blumea 23:417-438.

Di Fulvio, T. E. 1975. Estomatogenesis en Halophytum ameghinoi (Halophytaceae). Kurtziana 8:17-29.

Drenth, E. 1975. Taccaceae. Fl. Malesiana 7(4):806-819.

Dreyer, D. L. 1966. Citrus bitter principles--V. Botanical distribution and chemotaxonomy in the Rutaceae. Phytochemistry 5:367-378.

Eckardt, T. 1976. Classical morphological features of centrospermous families. Plant Syst. Ecol. 126:5-25.

Ehrendorfer, F. 1976a. Chromosome numbers and differentiation of centrospermous families. Plant Syst. Evol. 126:27-30.

------. 1976b. Closing remarks: systematics and evolution of centrospermous families. Plant Syst. Evol. 126:99-106.

Eiten, L. T. 1976. Inflorescence units in the Cyperaceae. Ann. Miss. Bot. Gard. 63:81-112.

Evans, F. J. and C. J. Soper. 1978. The tigliane, daphnane, and ingenane diterpenes, their chemistry, distribution and biological activities. A review. Lloydia 41:193-233.

Eyde, R. H. 1976. The bases of angiosperm phylogeny: floral anatomy. Ann. Miss. Bot. Gard. 62:521-537.

Fairbrothers, D. E., T. J. Mabry, R. L. Scogin and B. L. Turner. 1976. The bases of angiosperm phylogeny: chemotaxonomy. Ann. Miss. Bot. Gard. 62:765-800.

Ferguson, I. K. 1977. Cornaceae. World Pollen and Spore Flora 6:1-34.

Foster, A. S. 1959. The morphological and taxonomic significance of dichotomous venation in Kingdonia uniflora Balfour f. et W. W. Smith. Notes Roy. Bot. Gard. Edinburgh 23: 1-12.

------. 1961a. The floral morphology and relationships of Kingdonia uniflora. J. Arnold Arbor. 42:397-415.

------. 1961b. "The Phylogenetic Significance of Dichotomous Veination in Angiosperms". In Recent Adv. Bot. Univ. Toronto Press, pp. 971-975.

------. 1963. The morphology and relationships of Circaeaster. J. Arnold Arbor. 44:299-327.

------ and H. J. Arnott. 1960. Morphology and dichotomous vasculature of the leaf of Kingdonia uniflora. Amer. J. Bot. 47:684-698.

Gerdemann, J. W. 1968. Vesicular-arvuscular mycorrhiza and plant growth. Ann. Rev. Phytopath. 6:397-418.

Giannasi, D. E. and K. J. Niklas. 1977. Pakaraimoideae, Dipterocarpaceae of the Western hemisphere. IV. Phytochemistry. Taxon 26:380-385.

Gibson, A. C. 1973. Comparative anatomy of secondary xylem in Cactoideae (Cactaceae). Biotropica 5:29-65.

------. 1978. Rayless secondary xylem of *Halophytum*. Bull. Torrey Bot. Club 105:39-44.

------. 1979. Anatomy of *Koeberlinia* and *Canotia* revisited. Madrono 26:1-12.

------ and K. E. Horak. 1978. Systematic anatomy and phylogeny of Mexican columnar cacti. Ann. Miss. Bot. Gard. 65: 999-1057.

Goldblatt, P. 1978. A contribution to cytology in Cornales. Ann. Miss. Bot. Gard. 65:650-655.

------. 1979. Chromosome number in two cytologically unknown New World families, Tovariaceae and Vivianiaceae. Ann. Miss. Bot. Gard. 65:776-777.

Gornall, R. J., B. A. Bohm and R. Dahlgren. 1979. The distribution of flavonoids in the angiosperms. Bot. Notiser. 132:1-30.

Graham, S. A. 1975. Taxonomy of the Lythraceae in the southeastern United States. Sida 6:80-103.

------. 1978. The American species of *Nesaea* (Lythraceae) and their relationship to *Heimia* and *Decodon*. Syst. Bot. 2:61-71.

Hamann, U. 1975. Neue Untersuchungen zur Embryologie und Systematik der Centrolepidaceae. Bot. Jahrb. Syst. 96: 154-191.

------. 1976. Hydatellaceae--a new family of Monocotyledoneae. New Zealand J. Bot. 14:193-196.

Hammond, H. D. 1955. Systematic serological studies in Ranunculaceae. Serol. Mus. Bull. 14:1-3.

Hansen, B. 1975. Balanophoraceae. Fl. Malesiana 7(4):783-805.

Harborne, J. B. 1967. Comparative biochemistry of the flavonoids--IV. Correlations between chemistry, pollen morphology and systematics in the family Plumbaginaceae. Phytochem. 6:1415-1428.

------. 1969. Occurrence of flavonol 5-methyl ethers in higher plants and their systematic significance. Phytochem. 8: 419-423.

------ and C. A. Williams. 1973. A chemotaxonomic survey of flavonoids and simple phenols in leaves of the Ericaceae. Bot. J. Linnean Soc. 66:37-54.

Hegnauer, R. 1959. Die Verbreitung der Blausäure bei den Cormophyten. 3. Mitteiling die Blausäurehaltigen Gattungen. Pharm. Weekbl. 94:248-262.

------. 1960. Chemotaxonomische Betrachtungen. 10. Die systematische Bedeutung des Blausäuremerkmales. Sonder-Abdruck Pharm. Zentralhalle 99(6):322-329.

------. 1966. Aucubinartige glucoside. Uber ihre Verbreitung und Bedeutung als systematisches Merkmal. Pharm. Acta Helvet. 41:577-587.

------. 1969. "Chemical Evidence for the Classification of Some Plant Taxa." In Perspectives in Phytochemistry, edited by J. B. Harborne and T. Swain, pp. 121-138. Academic Press.

------. 1971. Pflanzenstoffe und Pflanzensystematik. Naturwissenschaften 58:585-598.

------. 1977. Cyanogenic compounds as systematic markers in Tracheophyta. Plant Syst. Evol., Suppl. 1:191-210.

------. 1978a. The importance of essential oils in plant classification. Dragoco Report 10:203-230.

------. 1978b. Phytochemie und Klassifikation der Umbelliferen, eine Neubewertung im Lichte der Seit 1972 Bekannt Gewordenen phytochemischen Tatsachen. Actes du 2e Sympos. Intern. Ombelliferes, Perpignan:335-363.

Heywood, V. H., J. B. Harborne and B. L. Turner (eds.). 1977. The Biology and Chemistry of the Compositae. 2 vols. New York: Academic Press.

Hickey, L. J. and J. A. Wolfe. 1976. The bases of angiosperm phylogeny: vegetative morphology. Ann. Miss. Bot. Gard. 62: 538-589.

Hunziker, J. H., H.-D. Behnke, I. J. Eifert and T. J. Mabry. 1974. *Halophytum ameghinoi:* a betalain-containing and P-type sieve-tube plastid species. Taxon 23:537-539.

Hutchinson, J. 1973. The Families of Flowering Plants Arranged According to a New System Based on Their Probable Phylogeny. 3rd Ed. Oxford: Clarendon Press.

Inamdar, J. A. and K. M. Aleykutty. 1979. Studies on *Cabomba aquatica* (Cabombaceae). Plant Syst. Evol. 132:161-166.

Jensen, S. R., B. J. Nielsen and R. Dahlgren. 1975. Iridoid compounds, their occurrence and systematic importance in the angiosperms. Bot. Notiser 128:148-180.

Jensen, U. 1968. Serologische Beitrage zur Systematik der Ranunculaceae. Bot. Jahrb. 88:269-310.

Johnson, L. A. S. and B. G. Briggs. 1975. On the Proteacae-- the evolution and classification of a southern family. Bot. J. Linnean Soc. 70:83-182.

Junell, S. 1931. Die Entwicklungsgeschichte von *Circaeaster agrestis*. Svensk. Bot. Tidskr. 25:238-270.

Keating, R. C. 1969. Comparative morphology of Cochlospermaceae. I. Synopsis of the family and wood anatomy. Phytomorph. 18:379-392

------. 1970. Comparative morphology of the Cochlospermaceae. II. Anatomy of the young vegetative shoot. Amer. J. Bot. 57:889-898.

------. 1972. The comparative morphology of the Cochlospermaceae. III. The flower and pollen. Ann. Miss. Bot. Gard. 59:282-296.

------. 1974. Trends of specialization in pollen of Flacourtiaceae with comparative observations of Cochlospermaceae and Bixaceae. Grana 15:29-49.

Keefe, J. M. and M. F. Moseley, Jr. 1978. Wood anatomy and phylogeny of *Paeonia* section *Moutan*. J. Arnold Arbor. 59: 274-297.

Kjaer, A. 1974. "The Natural Distribution of Glucosinolates:

A Uniform Group of Sulfur-Containing Glucosides." In Chemistry in Botanical Classification. edited by G. Bendz and J. Santesson, pp. 229-234. Stockholm: Nobel Found.

Kooiman, P. 1974. Iridoid glycosides in the Loasaceae and the taxonomic position of the family. Acta Bot. Neerl. 23: 677-679.

Kostermans, A. J. G. H. 1978. *Pakaraimaea dipterocarpacea* belongs to Tiliaceae and not to Dipterocarpaceae. Taxon 27: 357-359.

Kowal, R. R., S. A. Mori and J. A. Kallunki. 1977. Chromosome numbers of Panamanian Lecythidaceae and their use in subfamilial classification. Brittonia 29:399-410.

Kubitzki, K. 1969. Chemosystematische Betrachtungen zur Grossgliederung der Dicotylen. Taxon 18:360-368.

Kumazawa, M. 1938. Systematic and phylogenetic consideration of the Ranunculaceae and Berberidaceae. Bot. Mag. (Tokyo) 52:9-15, 52-53.

Kurosawa, S. 1977. Notes on chromosome numbers of spermatophytes. J. Jap. Bot. 52:225-230.

Lee, D. W. and D. E. Fairbrothers. 1972. Taxonomic placement of the Typhales within the monocotyledons: preliminary serological investigation. Taxon 21:39-44.

Lefor, M. W. 1975. A taxonomic revision of the Vivianiaceae. Univ. Conn. Occ. Pap. Biol. Sci. Ser. 2:225-255.

Mabry, T. J. "Is the Order Centrospermae Monophyletic? A Review of Phylogeneticalllly Significant, Molecular, Ultrastructural and Other Data for Centrospermous Families." In Chemistry in Botanical Classification, edited by G. Bendz and J. Santesson, pp. 275-280. Stockholm: Nobel Found.

Mabry, T. J. 1976. Pigment dichotomy and DNA-RNA hybridization data for centrospermous families. Plant Syst. Evol. 126:79-94.

------. 1977. The order Centrospermae. Ann. Miss. Bot. Gard. 64:210-220.

‾‾‾‾‾‾ and H.-D. Behnke (eds.). 1976. Evolution of centrospermous families. A symposium held on July 8, 1975 during the XIIth International Botanical Congress, Leningrad (USSR). Plant Syst. Evol. 126:1-106.

‾‾‾‾‾‾, P. Neuman and W. R. Philipson. 1978. *Hectorella*: a member of the betalain-suborder Chenopodiineae of the order Centrospermae. Plant Syst. Evol. 130:163-165.

Maguire, B., et al. 1977. Pakaraimoideae, Dipterocarpaceae of the western hemisphere. Taxon 26:341-385. I. Introduction. B. Maguire, p. 341.

Maguire, B. and P. S. Ashton. 1977. Pakaraimoideae, Dipterocarpaceae of the western hemisphere. II. Systematic, geographic and phyletic considerations. Taxon 26:343-368.

‾‾‾‾‾‾ and J. M. Pires. Saccifoliaceae. A new monotypic family of the Gentianales. Mem. N.Y. Bot. Gard. 29:230-245.

Marburger, J. E. 1979. Glandular leaf structure of *Triphyophyllum peltatum* (Dioncophyllaceae): a "fly paper" insect trapper. Amer. J. Bot. 66:404-411.

Maury, G., J. Muller and B. Lugardon. 1975. Notes on the morphology and fine structure of the exine of some pollen types in Dipterocarpaceae. Rev. Palaeobot. Palynol. 19: 241-289.

Meeuse, A. D. J. 1975. Taxonomic relationships of Salicaceae and Flacourtiaceae: their bearing on interpretative floral morphology and dilleniid phylogeny. Acta Bot. Neerl. 24: 437-457.

Melchior, H. (ed.). 1964. A. Engler's Syllabus der Pflanzenfamilien, II. Angiospermen. 12th Ed. Berlin: Gebr. Borntraeger.

Moore, H. E., Jr. 1973. The major groups of palms and their distribution. Gentes Herb. 11(2):27-141.

Mori, S. A., G. T. Prance and A. B. Bolten. 1978. Additional notes on the floral biology of neotropical Lecythidaceae. Brittonia 30:113-130.

Muller, J. 1975. Note on the pollen morphology of Crypteroniaceae s. l. Blumea 22:275-294.

Neubauer, H. F. 1978. Über Knotenbau und Vaskularisation von Blattgrund und Blattstiel bei einigen Cornaceae und einigen ihnen nahestehenden Arten, sowie uber Knotenbau im allgemeinen. Bot. Jahrb. Syst. 99:410-424.

Nooteboom, H. P. 1975. Revision of the Symplocaceae of the Old World (New Caledonia excepted). The Hague: Leiden Univ. Press.

Orchard, A. E. 1975. Taxonomic revisions in the family Haloragaceae. I. The genera *Haloragis*, *Haloragodendron*, *Glischrocaryon*, *Meziella* and *Gonocarpus*. Bull. Auckland Inst. & Mus. 10:1-299.

Ourisson, G. 1974. "Some Aspects of the Distribution of Diterpenes in Plants." In Chemistry in Botanical Classification, edited by G. Bendz and J. Santesson, pp. 129-134. Stockholm: Nobel Found.

Parmelee, J. A. and D. B. O. Savile. 1954. Life history and relationship of the rusts of *Sparganium* and *Acorus*. Mycologia 46:823-836.

Pennington, T. D. and B. T. Styles. 1975. A generic monograph of the Meliaceae. Blumea 22:419-540.

Petersen, F. and D. E. Fairbrothers. 1979. *Amphipterygium*--an amentiferous member of the Anacardiaceae. Abst. Phytochem. Bull. 12:28-29.

Posluszny, U. and P. B. Tomlinson. 1977. Morphology and development of floral shoots and organs in certain Zannichelliaceae. Bot. J. Linnean Soc. 75:21-46.

Poston, M. E. and H. J. Thompson. 1977. Cytotaxonomic observations in Loasaceae subfamily Loasoideae. Syst. Bot. 2:28-35.

Prance, G. T. 1976. The pollination and androphore structure of some Amazonian Lecythidaceae. Biotropica 4:235-241.

------ and S. A. Mori. 1977. What is *Lecythis*? Taxon 26:209-222.

------ and ------. 1978. Observations on the fruits and seeds of neotropical Lecythidaceae. Brittonia 30:21-33.

Puff, C. and A. Weber. 1976. Contributions to the morphology, anatomy, and karyology of *Rhabdodendron*, and a reconsideration of the systematic position of the Rhabdodendraceae. Plant Syst. Evol. 125:195-222.

Raven, P. H. 1976. The bases of angiosperm phylogeny: cytology. Ann. Miss. Bot. Gard. 62:724-764.

Rezende, C. M. A. M., O. R. Gottlieb and M. C. Marx. 1975. Benzyltetrahydroisoquinoline-derived alkaloids as systematic markers. Biochem. System. Ecol. 3:63-70.

Richardson, M. 1978. Flavonols and C-glycosylflavonoids of the Caryophyllales. Biochem. System. Ecol. 6:283-286.

Rodman, J. E. 1981. "Glucosinolates: Divergence and Convergence in Phytochemical Characters." In Phytochemistry and Angiosperm Phylogeny, edited by D. A. Young and D. S. Seigler. (in press).

Rodriguez, R. L. 1971. "The Relationships of the Umbellales." In The Biology and Chemistry of the Umbelliferae, edited by V. H. Heywood, pp. 63-92. Suppl. 1, Bot. J. Linnean Soc. 64. London: Academic Press.

Roth, I. 1977. Fruits of Angiosperms. Encyclopedia of Plant Anatomy 10(1):1-675. Berlin: Borntraeger.

Saupe, S. G. 1981. "Cyanogenesis and Angiosperm Phylogeny." In Phytochemistry and Angiosperm Phylogeny, edited by D. A. Young and D. S. Seigler (in press).

Sawada, M. 1971. Floral vascularization of *Paeonia japonica* with some consideration on systematic position of the Paeoniaceae. Bot. Mag. (Tokyo) 84:51-60.

Schmid, R. 1979a. Reproductive anatomy of *Actinidia chinensis* (Actinidiaceae). Bot. Jahrb. Syst. 100:149-195.

------. 1979b. Actinidiaceae, Davidiaceae, and Paracryphiaceae: systematic considerations. Bot. Jahrb. Syst. 100:196-204.

------. 1980. Comparative anatomy and morphology of *Psiloxylon* and *Heteropyxis*, and the subfamilial and tribal classification of Myrtaceae. Taxon 29:559-595.

Scogin, R. 1977. Anthocyanins of the Fouquieriaceae. Biochem.

System. Ecol. 5:265-267.

------. 1978. Leaf phenolics of the Fouquieriaceae. Biochem. System. Ecol. 6:297-298.

Seigler, D. S. 1977. Plant systematics and alkaloids. The Alkaloids 16:1-82. New York: Academic Press.

------. 1981. "Terpenoids and Plant Systematics." In Phytochemistry and Angiosperm Phylogeny, edited by D. A. Young and D. S. Seigler. (in press).

Shen, Y.-F. 1954. Phylogeny and wood anatomy of Nandina. Taiwania No. 5:85-92.

Simon, J.-P. 1970. Comparative serology of the order Nymphaeales. I. Preliminary survey on the relationships of Nelumbo. Aliso 7:243-261.

------. 1971. Comparative serology of the order Nymphaeales. II. Relationships of Nymphaeaceae and Nelumbonaceae. Aliso 7:325-250.

Skvarla, J. M. and J. W. Nowicke. 1976. Ultrastructure of pollen exine in centrospermous families. Plant Syst. Ecol. 126:55-78.

------, P. H. Raven and J. Praglowski. 1975. The evolution of pollen tetrads in Onagraceae. Amer. J. Bot. 62:6-35.

-----, ----- and ------. 1976. "Ultrastructural Survey of Onagraceae Pollen." In The Evolutionary Significance of the Exine, edited by I. K. Ferguson and J. Muller. Linnean Soc. Symp. Ser. 1:447-479.

-----, -----, W. F. Chissoe and M. Sharp. 1978. An ultrastructural study of viscin threads in Onagraceae pollen. Pollen et Spores 20:5-143.

Smith, L. B. 1974. Bromeliaceae subfamily Pitcairnioideae. Flora Neotropica 14:1-660.

------ and E. S. Ayensu. 1974. Classification of Old World Velloziaceae. Kew Bull. 29:181-205.

------ and ------. 1976. A revision of American Velloziaceae. Smithsonian Contr. Bot. 30:1-172.

Spongberg, S. A. 1976. Styracaceae hardy in temperate North America. J. Arnold Arbor. 57:54-73.

Stafleu, F. A. (Chairman, Ed. Comm.), et al. 1978. International Code of Botanical Nomenclature. Reg. Veg. 97:1-457.

Stevens, P. F. 1971. A classification of the Ericaceae: subfamilies and tribes. Bot. J. Linnean Soc. 64:1-53.

Takhtajan, A. 1959. Die Evolution der Angiospermen. Transl. from the Russian by W. Hoppner, Berlin. Jena: G. Fischer.

------. 1966. Systema et Phylogenia Magnoliphytorum (in Russian). Inst. Bot. Komarov, Acad. Sci. USSR, Moscow-Leningrad.

------. 1969. Flowering Plants: Origin and Dispersal. Transl. from the Russian by C. Jeffrey, Kew. Edinburgh: Oliver and Boyd.

------. 1974. "The Chemical Approach to Plant Classification with Special Reference to the Higher Taxa of Magnoliophyta." In Chemistry in Botanical Classification, edited by G. Bendz and J. Santesson, pp. 17-26. Uppsala: Nobel Found.

Tamura, M. 1972. Morphology and phyletic relatinship of the Glaucidiaceae. Bot. Mag. Tokyo 85:29-41.

Taylor, N. 1909. Cymodoceaceae. North Amer. Flora 17(1): 31-32.

Thompson, J. 1976. A revision of the genus *Tetratheca* (Tremandraceae). Telopea 1:139-215.

Thorne, R. F. 1968. Synopsis of a putatively phylogenetic classification of the flowering plants. Aliso 6(4):57-66.

------. 1974a. The "Amentiferae" or Hamamelidae as an artificial group: a summary statement. Brittonia 25:395-405.

------. 1974b. A phylogenetic classification of the Annoniflorae. Aliso 8:147-209.

------. 1976. A phylogenetic classification of the Angiospermae. Evol. Biol. 9:35-106.

------. 1977. Some realignments in the Angiospermae. Plant Syst. Evol., Suppl. 1:299-319.

------. 1979. "Parasites and Phytophages--Pragmatic Chemists?" In Parasites and Phylogeny, edited by O. Hedberg and I. Hedberg. Spec. Publ. Univ. Uppsala 22:200-209.

------ and R. Scogin. 1978. *Forsellesia* Greene (*Glossopetalon* Gray), a third genus in the Crossosomataceae, Rosineae, Rosales. Aliso 9:171-178.

Tomlinson, P. B. 1969. Anatomy of the Monocotyledons, edited by C. A. Metcalfe. III. Commelinales-Zingiberales. Oxford: Clarendon Press.

------ and U. Posluszny. 1976. Generic limits in the Zannichelliaceae (sensu Dumortier). Taxon 25:273-279.

------ and A. C. Smith. 1970. Joinvilleaceae, a new family of monocotyledons. Taxon 19:887-889.

Uhl, N. W. and H. E. Moore, Jr. 1977. Centrifugal stamen initiation in Phytelephantoid palms. Amer. J. Bot. 64:1152-1161.

Valen, F. van. 1978. Contribution to the knowledge of cyanogenesis in Angiosperms. 9. Communication. Cyanogenesis in Papaverales. Proc. Koninklijke Ned. Akad. Wetenschappen, Amsterdam, Ser. C, 81:492-499.

Vijayaraghavan, M. R. and A. Vidya Kumari. 1974. Embryology and systematic position of *Zannichellia palustris* L. J. Indian Bot. Soc. 53:292-302.

Vliet, G. J. C. M. van. 1974. Wood anatomy of Crypteroniaceae sensu lato. J. Microscopy 104:65-82.

------ and P. Baas. 1975. Comparative anatomy of the Crypteroniaceae sensu lato. Blumea 22:175-195.

Walker, J. W. 1976. "Comparative Pollen Morphology and Phylogeny of the Ranalean Complex." In Origin and Early Evolution of Angiosperms, edited by C. B. Beck, pp. 241-299. New York: Columbia Univ. Press.

Wallace, G. D. 1975. Studies of the Monotropoideae (Ericaceae): taxonomy and distribution. Wasmann J. Biol. 33:1-88.

------. 1976. Interrelationships of the subfamilies of the Ericaceae and derivations of the Monotropoideae. Bot. Notiser. 128:286-298.

------. 1977. Studies of the Monotropoideae (Ericaceae). Floral nectaries: anatomy and function in pollination ecology. Amer. J. Bot. 64:199-206.

Waterman, P. G. 1975. Alkaloids of the Rutaceae: their distribution and systematic significance. Biochem. Syst. Ecol. 3:149-180.

Watson, L. 1964. The taxonomic significance of certain anatomical observations on Ericaceae. New Phytologist 63:274-280.

------. 1965. The taxonomic significance of certain anatomical variations among Ericaceae. J. Linnean Soc., Bot. 59:111-125.

Whitmore, T. C. 1974. *Abdulmajidia*, a new genus of Lecythidaceae from Malaysia. Kew Bull. 29:207-211.

Wunderlich, R. 1971. Die systematische Stellung von *Theligonum*. (Zugleich eine kritische Zusammenstellung einiger embryologischer, anatomischer und morphologischer Merkmale der Rubiaceae). Osterr. Bot. Z. 119:329-394.

Yoong, Ng Siew, W. R. Philipson and J. R. L. Walker. 1975. Hectorellaceae--a member of the Centrospermae. New Zealand J. Bot. 13:567-570.

Young, D. A. 1976. Flavonoid chemistry and the phylogenetic relationships of the Julianiaceae. Syst. Bot. 1:149-162.

------. 1981. "The Usefulness of Flavonoids in Angiosperm Phylogeny: Some Selected Examples." In Phytochemistry and Angiosperm Phylogeny, edited by D. A. Young and D. S. Seigler. (in press).

Zeeuw, C. de. 1977. Pakaraimoideae, Dipterocarpaceae of the western hemisphere. III. Stem anatomy. Taxon 26:368-380.

A SYNOPSIS OF THE CLASS ANGIOSPERMAE (ANNONOPSIDA)

Subclass: DICOTYLEDONEAE (ANNONIDAE)

Superorder: Annoniflorae (C; 458/12,570)

Order: Annonales (C; 279/9,020)
  Suborder: Winterineae (C; 8/90)
    Family: Winteraceae (C; 8/90)
  Suborder: Illiciineae (C; 3/84)
    Family: Illiciaceae (C; 1/37)
    Family: Schisandraceae (C/ 2/47)
  Suborder: Annonineae (incl. Magnoliineae) (C; 170/2,850)
    Family: Magnoliaceae (C; 12/220)
      Subfamily: Magnoliodeae (11/220)
      Subfamily: Liriodendroideae (1/2)
    Family: Degeneriaceae (C; 1/1)
    Family: Himantandraceae (C; 1/1)
    Family: Eupomatiaceae (C; 1/2)
    Family: Annonaceae (C; 132/2300)
      Subfamily: Fusaeoideae
      Subfamily: Annonoideae
      Subfamily: Malmeoideae
    Family: Myristicaceae (C; 17/300)
    Family: Canellaceae (C; 6/21)
  Suborder: Aristolochiineae (C; 7/400)
    Family: Aristolochiaceae (C; 7/400)
  Suborder: Laurineae (C; 78/2, 490)
    Family: Amborellaceae (C; 1/1)
    Family: Austrobaileyaceae (C; 1/2)
    Family: Trimeniaceae (B; 2/5)
    Family: Chloranthaceae (C; 5/70)
    Family: Lactoridaceae (C; 1/1)
    Family: Monimiaceae (C; 29/327)
      Subfamily: Hortonioideae (C; 1/2)
      Subfamily: Monimioideae (C; 2/5)
      Subfamily: Mollinedioideae (C; 18/145)
      Subfamily: Atherospermatoideae (C; 5/12)
      Subfamily: Siparunoideae (C; 2/160)
      Subfamily: Glossocalycoideae (C; 1/3)
    Family: Gomortegaceae (C; 1/1)
    Family: Calycanthaceae (C; 3/7)
      Subfamily: Idiospermoideae (C; 1/1)
      Subfamily: Calycanthoideae (C; 2/6)
    Family: Lauraceae (C; 31/2,020)
      Subfamily: Lauroideae (C; 30/2,000)
      Subfamily: Cassythoideae (C; 1/20)
    Family: Hernandiaceae (C; 4/58)
      Subfamily: Gyrocarpoideae (C; 2/16)
      Subfamily: Hernandioideae (C; 2/42)
  Suborder: Piperineae (C; 13/3,110)
    Family: Saururaceae (C; 5/7)
    Family: Piperaceae (C; 8/3,100)
      Subfamily: Piperoideae (C; 4/2,100)
      Subfamily: Peperomioideae (C; 4/1,000)

Order: Nelumbonales (C; 1/2)
    Family: Nelumbonaceae (C; 1/2)
Order: Paeoniales (B; 1/33)
    Family: Paeoniaceae (C; 1/33)
    Family: Glaucidiaceae (C; 1/1)
Order: Berberidales (C; 177/3,520)
    Suborder: Berberidineae (C; 135/2,850)
        Family: Lardizabalaceae (C; 8/35)
        Family: Sargentodoxaceae (C; 1/1)
        Family: Menispermaceae (C; 65/350)
        Family: Nandinaceae (C; 1/1)
        Family: Berberidaceae (C; 12560)
            Subfamily: Berberidoideae (3/520)
            Subfamily: Epimedioideae (4/28)
            Subfamily: Leonticoideae (3/6)
            Subfamily: Podophylloideae (2/9)
        Family: Ranunculaceae (C; 46/1,900)
            Subfamily: Hydrastidoideae (1/2)
            Subfamily: Thalictroideae
            Subfamily: Ranunculoideae
        Family: Circaeasteraceae (C; 2/2)
            Subfamily: Kingdonioideae (1/1)
            Subfamily: Circaeasteroideae (1/1)
    Suborder: Papaverineae (C; 42/665)
        Family: Papaveraceae (C; 42/665)
            Subfamily: Platystemonoideae (3/?)
            Subfamily: Papaveroideae (8/?)
            Subfamily: Chelidonioideae (10/?)
            Subfamily: Eschscholzioideae (3/?)
            Subfamily: Pteridophylloideae (1/1)
            Subfamily: Hypecoideae (1/15)
            Subfamily: Fumarioideae (16/450)

Superorder: Nymphaeiflorae (C; 9/75)

Order: Nymphaeales (C; 9/75)
    Family: Cabombaceae (C; 2/8)
    Family: Nymphaeaceae (C; 6/62)
        Subfamily: Nymphaeoideae (3/55)
        Subfamily: Euryaloideae (2/3)
        Subfamily: Barclayoideae (1/4)
    Family: Ceratophyllaceae (C; 1/6)

Superorder: Rafflesiiflorae (A; 10/67)

Order: Rafflesiales (B; 10/67)
    Family: Rafflesiaceae (C; 8/53)
        Subfamily: Cytinoideae (2/10)
        Subfamily: Apodanthoideae (2/26)
        Subfamily: Rafflesioideae (3/15)
        Subfamily: Mitrastemonoideae (1/2)
    Family: Hydnoraceae (B; 2/14)

Superorder: Theiflorae (C; 575/12,800)

Order: Theales (C; 257/4,950)
   Suborder: Dilleniineae (C; 9/400)
      Family: Dilleniaceae (C; 9/400)
         Subfamily: Tetraceroideae (4/120)
         Subfamily: Dillenioideae (5/280)
   Suborder: Theineae (C; 114/2,240)
      Family: Actinidiaceae (C; 3/350
         Subfamily: Actinidioideae (1/40)
         Subfamily: Saurauioideae (1/300)
         Subfamily: Clematoclethroideae (1/10)
      Family: Paracryphiaceae (B; 1/1)
      Family: Stachyuraceae (C; 1/10)
      Family: Theaceae (C; 39/540)
         Subfamily: Ternstroemioideae (12/200)
         Subfamily: Theoideae (16/300)
         Subfamily: Bonnetioideae (7/29)
         Subfamily: Asteropeioideae (1/7)
         Subfamily: Tetrameristoideae (2/2)
         Subfamily: Pellicieroideae (1/1)
      Family: Symplocaceae (C; 1/500)
      Family: Caryocaraceae (C; 2/23)
      Family: Oncothecaceae (B; 1/1)
      Family: Aquifoliaceae (C; 2/400)
      Family: Phellinaceae (C; 1/10)
      Family: Icacinaceae (B; 56/300)
      Family: Sphenostemonaceae (B; 1/7)
      Family: Cardiopteridaceae (B/ 1/3)
      Family: Marcgraviaceae (C; 5/100)
   Suborder: Clethrineae (C; 5/80)
      Family: Pentaphylacaceae (C; 1/2)
      Family: Clethraceae (C; 1/64)
      Family: Cyrillaceae (C; 3/14)
   Suborder: Sarraceniineae (C; 3/15)
      Family: Sarraceniaceae (C; 3/15)
   Suborder: Scytopetalineae (C; 57/710)
      Family: Ochnaceae (C; 40/600)
         Subfamily: Ochnoideae
         Subfamily: Diegodendroideae (1/1)
         Subfamily: Sauvagesioideae
      Family: Quiinaceae (C; 4/50)
      Family: Scytopetalaceae (C; 5/20)
      Family: Sphaerosepalaceae (B; 2/14)
      Family: Medusagynaceae (C; 1/1)
      Family: Strasburgeriaceae (B; 1/1)
      Family: Ancistrocladaceae (B; 1/20)
      Family: Dioncophyllaceae (B; 3/3)
   Suborder: Nepenthineae (C; 1, 70)
      Family: Nepenthaceae (C; 1/70)
   Suborder: Hypericineae (C; 43/1,040)
      Family: Clusiaceae (incl. Hypericaceae) (C; 41/1,000)
         Subfamily: Kielmeyeroideae (B; 3/40)
         Subfamily: Calophylloideae
         Subfamily: Clusioideae
         Subfamily: Moronoboideae (6/36)
         Subfamily: Hypericoideae (9/540)
      Family: Elatinaceae (C; 2/40)
   Suborder: Lecythidineae (C; 25/400)
      Family: Lecythidaceae (C; 25/400)

        Subfamily: Planchonioideae (incl. *Abdulmajidia*) (6/54)
        Subfamily: Foetidioideae (1/5)
        Subfamily: Lecythidoideae (15/325)
        Subfamily: Napeolonaeoideae (2/18)
        Subfamily: Asteranthoideae (1/1)
Order: Ericales (C; 130/2,420)
    Family: Ericaceae (C; 98/2,010)
        Subfamily: Rhododendroideae (15/700)
        Subfamily: Ericoideae (16/600)
        Subfamily: Vacciniodeae (incl. Arbutoideae) (54/660)
        Subfamily: Pyroloideae (3/30)
        Subfamily: Monotropoideae (10/12)
    Family: Epacridaceae (C; 31/400)
        Subfamily: Epacridoideae (30/400)
        Subfamily: Wittsteinioideae (1/1)
    Family: Empetraceae (C; 3/6)
Order: Ebenales (C; 66/1,450)
   Suborder: Ebenineae (Sapotineae) (C; 53/1,300)
     Family: Ebenaceae (incl. *Lissocarpa*) (C; 3/500)
     Family: Sapotaceae (incl. *Sarcosperma*) (C; 50/800)
   Suborder: Styracineae (C; 13/150)
     Family: Styracaceae (C; 13/150)
Order: Primulales (C; 77/2,880)
   Suborder: Primulineae (C; 58/2,110)
     Family: Myrsinaceae (C; 38/1,110)
        Subfamily: Theophrastoideae (5/110)
        Subfamily: Myrsinoideae (incl. *Aegiceras*) (32/900)
        Subfamily: Maesoideae (1/100)
     Family: Primulaceae (C; 20/1,000)
   Suborder: Plumbaginineae (C; 19/770)
     Family: Plumbaginaceae (C; 19/770)
        Subfamily: Plumbaginoideae (4/24)
        Subfamily: Staticoideae (incl. *Aegialitis*) (15/750)
Order: Polygonales (A; 45/1,100)
    Family: Polygonaceae (45/1,100)
        Subfamily: Eriogonoideae (13/320)
        Subfamily: Polygonoideae (24/550)
        Subfamily: Coccoloboideae (8/230)

Superorder: Chenopodiiflorae (Centrospermae) (C; 550/8,610)

  Order: Chenopodiales (C; 550/8,610)
   Suborder: Chenopodiineae (incl. Phytolaccineae and Caryophyllineae)
                                                       (C; 416/6,450)
     Family: Phytolaccaceae (C; 18/120)
        Subfamily: Phytolaccoideae (4/45)
        Subfamily: Gisekioideae (1/5)
        Subfamily: Rivinoideae (6/40)
        Subfamily: Agdestidoideae (1/1)
        Subfamily: Microteoideae (2/15)
        Subfamily: Stegnospermatoideae (1/3)
        Subfamily: Barbeuioideae (1/1)
        Subfamily: Achatocarpoideae (2/10)
     Family: Aizoaceae (C; 130/2,040)
        Subfamily: Aizooideae (10/50)
        Subfamily: Mesembryanthemoideae (Aptenioideae) (3/47)
        Subfamily: Hymenogynoideae (1/3)

        Subfamily: Caryotophoroideae (1/1)
        Subfamily: Ruschioideae (100/1,800)
        Subfamily: Tetragonioideae (1/50)
        Subfamily: Molluginoideae (13/90)
     Family: Caryophyllaceae (incl. *Geocarpon*) (C; 70/1,750)
        Subfamily: Alsinoideae
        Subfamily: Paronychioideae
        Subfamily: Caryophylloideae
     Family: Halophytaceae (C; 1/1)
     Family: Nyctaginaceae (C; 30/290)
     Family: Chenopodiaceae (C; 102/1,400)
        Subfamily: Chenopodioideae (incl. *Dysphania*)
        Subfamily: Salicornioideae
        Subfamily: Salsoloideae
     Family: Amaranthaceae (C; 65/850)
        Subfamily: Amaranthoideae
        Subfamily: Gomphrenoideae
  Suborder: Portulacineae (C; 134/2,160)
     Family: Portulacaceae (C; 21/580)
        Subfamily: Portulacoideae (19/575)
        Subfamily: Hectorelloideae (2/2)
     Family: Basellaceae (C; 4/20)
     Family: Didiereaceae (C; 4/11)
     Family: Cactaceae (C; 105/1,550)
        Subfamily: Pereskioideae (2/13)
        Subfamily: Opuntioideae (8/240)
        Subfamily: Cactoideae (95/1,300)

Superorder: Geraniiflorae (B; 172/5,100)

  Order: Geraniales (C; 172/5,100)
     Suborder: Linineae (C; 57/880)
        Family: Houmiriaceae (C; 8/50)
        Family: Ctenolophonaceae (C; 1/3)
        Family: Linaceae (C; 20/334)
           Subfamily: Linoideae (12/290)
           Subfamily: Ixonanthoideae (8/44)
        Family: Erythroxylaceae (C; 2/250)
        Family: Zygophyllaceae (C; 26/240)
           Subfamily: Peganoideae (2/6)
           Subfamily: Morkillioideae (Chitonioideae) (2/4)
           Subfamily: Tetradiclidoideae (1/1)
           Subfamily: Tribuloideae (4/45)
           Subfamily: Zygophylloideae (incl. Augeoideae) (16/176)
           Subfamily: Nitrarioideae (1/8)
     Suborder: Geraniineae (C; 29/2,370)
        Family: Oxalidaceae (incl. Averrhoaceae, *Lepidobotrys*) (C; 7/890)
        Family: Geraniaceae (C; 7/760)
           Subfamily: Geranioideae (5/750)
           Subfamily: Dirachmoideae (1/1)
           Subfamily: Biebersteinioideae (1/5)
        Family: Vivianiaceae (B; 4/6)
        Family: Ledocarpaceae (B; 2/11)
        Family: Balsaminaceae (C; 5/600)
        Family: Tropaeolaceae (C; 3/92)
        Family: Limnanthaceae (C; 1/11)

    Suborder: Polygalineae (C; 86/1,850)
        Family: Malpighiaceae (C; 69/800)
            Subfamily: Gaudichaudioideae
            Subfamily: Malpighioideae
            Subfamily: Byrsonimoideae
        Family: Polygalaceae (incl. *Diclidanthera, Xanthophyllum*) (C; 15/800)
        Family: Krameriaceae (C; 1/25)
        Family: Trigoniaceae (C; 3/26)
        Family: Vochysiaceae (C; 7/200)

Superorder: Santaliflorae (B; 221/2,960)

  Order: Celastrales (B; 61/890)
        Family: Celastraceae (incl. *Canotia*, excl. *Forsellesia*) (C; 57/860)
            Subfamily: Celastroideae (25/?)
            Subfamily: Tripterygioideae (6/?)
            Subfamily: Cassinoideae (20/?)
            Subfamily: Hippocrateoideae (4/300)
            Subfamily: Goupioideae (1/3)
            Subfamily: Siphonodontoideae (1/5)
        Family: Lophopyxidaceae (B; 1/2)
        Family: Stackhousiaceae (B; 3/27)
            Subfamily: Macgregorioideae (1/1)
            Subfamily: Stackhousioideae (2/26)
  Order: Santalales (C; 141/2,020)
        Family: Olacaceae (C; 33/310)
            Subfamily: Schoepfioideae (incl. *Octoknema*)
            Subfamily: Olacoideae
            Subfamily: Opilioideae (8/60)
        Family: Medusandraceae (C; 1/2)
        Family: Santalaceae (incl. *Okoubaka*) (C; 30/400)
        Family: Eremolepidaceae (C; 3/11)
        Family: Misodendraceae (C; 1/11)
        Family: Loranthaceae (C; 65/850)
        Family: Viscaceae (C 8/440)
  Order: Balanophorales (B; 19/45)
        Family: Balanophoraceae (C; 18/45)
            Subfamily: Mystropetaloideae (1/1)
            Subfamily: Dactylanthoideae (2/2)
            Subfamily: Sarcophytoideae (2/3)
            Subfamily: Helosidoideae (6/12)
            Subfamily: Lophophytoideae (4/8)
            Subfamily: Balanophoroideae (3/19)
        Family: Cynomoriaceae (B; 1/2)

Superorder: Violiflorae (C; 722/9,820)

  Order: Violales (C; 292/5,610)
    Suborder: Violineae (incl. Caricineae) (C; 150/3,030)
        Family: Flacourtiaceae (incl. *Lacistema, Neopringlea, Prockia*)
                                                    (C; 91/1,280)
        Family: Dipentodontaceae (C; 1/1)
        Family: Peridiscaceae (B; 2/2)
        Family: Scyphostegiaceae (C; 1/1)
        Family: Violaceae (C; 22/900)

   Subfamily: Violoideae (21/894)
   Subfamily: Leonioideae (1/6)
  Family: Passifloraceae (incl. Paropsieae) (C; 18/630)
  Family: Turneraceae (C; 8/120)
  Family: Malesherbiaceae (C; 1/35)
  Family: Achariaceae (C; 3/3)
  Family: Caricaceae (C; 4/55)
 Suborder: Salicineae (C; 2/530)
  Family: Salicaceae (C; 2/530)
 Suborder: Tamaricineae (C; 9/210)
  Family: Tamaricaceae (incl. *Tamaricaria*) (C; 5/120)
  Family: Frankeniaceae (C; 4/90)
 Suborder: Cucurbitineae (C; 110/640)
  Family: Cucurbitaceae (C; 110/640)
   Subfamily: Cucurbitoideae
   Subfamily: Zanonioideae
 Suborder: Begoniineae (C; 8/924)
  Family: Begoniaceae (C; 5/920)
  Family: Datiscaceae (incl. Tetrameleae) (C; 3/4)
 Suborder: Loasasineae (A; 13/280)
  Family: Loasaceae (C; 13/280)
   Subfamily: Mentzelioideae
   Subfamily: Loasoideae
   Subfamily: Gronovioideae
Order: Capparales (C; 429/4,210)
  Family: Moringaceae (B; 1/12)
  Family: Resedaceae (C; 6/70)
  Family: Capparaceae (incl. *Oceanopapaver*) (C; 46/930)
   Subfamily: Pentadiplandroideae (B; 1/2)
   Subfamily: Tovarioideae (C; 1/2)
   Subfamily: Koeberlinioideae (C; 1/1)
   Subfamily: Capparoideae (C; 30/640)
   Subfamily: Cleomoideae (incl. *Buhsia*, *Podandrogyne*) (C; 13/285)
  Family: Brassicaceae (incl. *Dipterygium*) (C; 376/3,200)

Superorder: Malviflorae (C; 774/14,750)

 Order: Malvales (C; 250/3,520)
  Family: Sterculiaceae (incl. *Maxwellia*) (C; 69/700)
   Subfamily: Sterculioideae (12/?)
   Subfamily: Byttnerioideae (48/?)
  Family: Huaceae (incl. *Afrostyrax*) (C; 2/3)
  Family: Elaeocarpaceae (excl. *Muntingia*) (B; 9/350)
  Family: Plagiopteraceae (B; 1/1)
  Family: Tiliaceae (incl. *Muntingia, Goethalsia*) (C; 48/450)
   Subfamily: Brownlowioideae (12/56)
   Subfamily: Tetralicoideae (1/3)
   Subfamily: Tilioideae (35/390)
   Subfamily: Neotessmannioideae (1/1)
  Family: Dipterocarpaceae (C; 15/570)
   Subfamily: Dipterocarpoideae (16/550)
   Subfamily: Monotoideae (2/20)
   Subfamily: Pakaramaeoideae (1/1)
  Family: Sarcolaenaceae (C; 8/40)
  Family: Bixaceae (C; 1/4)
  Family: Cochlospermaceae (C; 2/20)

            Family: Cistaceae (C; 8/200)
            Family: Bombacaceae (C; 20/180)
            Family: Malvaceae (C; 75/1,000)
    Order: Urticales (excl. *Barbeya*) (C; 115/2,600)
            Family: Ulmaceae (C; 15/200)
               Subfamily: Ulmoideae
               Subfamily: Celtidoideae
            Family: Urticaceae (C; 98/2,400)
               Subfamily: Moroideae (53/1,400)
               Subfamily: Cecropioideae (6/200)
               Subfamily: Urticoideae (39/800)
            Family: Cannabaceae (C; 2/3)
    Order: Rhamnales (C; 48/900)
            Family: Rhamnaceae (C; 45/850)
            Family: Elaeagnaceae (C; 3/50)
    Order: Euphorbiales (C; 361/7,730)
            Family: Euphorbiaceae (C; 300/7,000)
               Subfamily: Phyllanthoideae
               Subfamily: Oldfieldioideae (incl. *Picrodendron*)
               Subfamily: Acalyphoideae
               Subfamily: Crotonoideae
               Subfamily: Euphorbioideae
            Family: Pandaceae (C; 4/28)
            Family: Simmondsiaceae (C; 1/1)
            Family: Aextoxicaceae (B; 1/1)
            Family: Didymelaceae (B; 1/2)
            Family: Dichapetalaceae (C; 4/200)
            Family: Thymelaeaceae (C; 50/500)
               Subfamily: Gonystyloideae (3/23)
               Subfamily: Aquilarioideae (5/40)
               Subfamily: Gilgiodaphnoideae (1/1)
               Subfamily: Thymelaeoideae (41/436)

Superorder: Rutiflorae (C; 1,120/19,870)

    Order: Rutales (C; 1,120/19,870)
       Suborder: Rutineae (C; 331/3,680)
            Family: Rutaceae (C; 154/920)
               Subfamily: Rutoideae
               Subfamily: Toddalioideae
               Subfamily: Dictylomatoideae (1/2)
               Subfamily: Flindersioideae (2/17)
               Subfamily: Spathelioideae (3/19)
               Subfamily: Aurantioideae
               Subfamily: Rhabdodendroideae (1/4)
            Family: Cneoraceae (C; 1/3)
            Family: Coriariaceae (B; 1/5)
            Family: Simaroubaceae (C; 33/250)
               Subfamily: Simarouboideae (21/120)
               Subfamily: Kirkioideae (1/8)
               Subfamily: Irvingioideae (8/48)
               Subfamily: Balanitoideae (1/25)
               Subfamily: Picramnioideae (1/40)
               Subfamily: Alvaradoideae (1/5)
            Family: Ptaeroxylaceae (C; 2/5)
            Family: Meliaceae (C; 53/1,310)
               Subfamily: Melioideae (incl. *Nymania*) (37/1,260)

            Subfamily: Quivisianthoideae (1/1)
            Subfamily: Capuronianthoideae (1/1)
            Subfamily: Neomangenotoideae (1/1)
            Subfamily: Swietenioideae (13/47)
        Family: Burseraceae (C; 16/500)
        Family: Anacardiaceae (incl. *Blepharocarya*, Julianiaceae) (C; 70/600)
        Family: Leitneriaceae (B; 1/1)
    Suborder: Juglandineae (C; 8/60)
        Family: Rhoipteleaceae (C; 1/1)
        Family: Juglandaceae (C; 7/60)
            Subfamily: Platycaryoideae (1/1)
            Subfamily: Juglandoideae (6/58)
    Suborder: Myricineae (C; 4/40)
        Family: Myricaceae (incl. *Canacomyrica*) (C; 4/40)
    Suborder: Sapindineae (C; 160/2,410)
        Family: Sapindaceae (incl. *Filicium*) (C; 143/2,000)
            Subfamily: Dodonaeoideae (29/120)
            Subfamily: Stylobasioideae (1/2)
            Subfamily: Emblingioideae (1/1)
            Subfamily: Sapindoideae (112/1,880)
        Family: Gyrostemonaceae (C; 5/16)
        Family: Bataceae (C; 1/2)
        Family: Sabiaceae (C; 3/160)
            Subfamily: Meliosmoideae (2/105)
            Subfamily: Sabioideae (B; 1/55)
        Family: Melianthaceae (C; 2/15)
        Family: Akaniaceae (C; 1/1)
        Family: Aceraceae (C; 2/200)
        Family: Hippocastanaceae (C; 2/15)
        Family: Bretschneideraceae (C; 1/2)
    Suborder: Fabineae (C; 617/13,680)
        Family: Surianaceae (C; 1/1)
        Family: Connaraceae (C; 16/325)
            Subfamily: Connaroideae (15/322)
            Subfamily: Jollydoroideae (1/3)
        Family: Fabaceae (C; 630/18,000)
            Subfamily: Caesalpinoideae (150/2,700)
            Subfamily: Mimosoideae (40/2,500)
            Subfamily: Faboideae (incl. Swartzieae) (440/12,800)

Superorder: Proteiflorae (B; 75/1,050)

    Order: Proteales (C; 74/1,050)
        Family: Proteaceae (C; 74/1,050)
            Subfamily: Persoonioideae (7/?)
            Subfamily: Proteoideae (26/?)
            Subfamily: Sphalmioideae (1/1)
            Subfamily: Carnarvonioideae (1/2)
            Subfamily: Grevilleoideae (40/?)

Superorder: Hamamelidiflorae (C; 50/1,410)

    Order: Hamamelidales (C; 34/110)
        Suborder: Trochodendrineae (C; 4/6)
            Family: Trochodendraceae (C; 1/1)

            Family: Tetracentraceae (C; 1/1)
            Family: Eupteleaceae (C; 1/2)
            Family: Cercidiphyllaceae (C; 1/2)
        Suborder: Eucommineae (C; 1/1)
            Family: Eucommiaceae (C; 1/1)
        Suborder: Hamamelidineae (C; 29/100)
            Family: Hamamelidaceae (C; 28/90)
                Subfamily: Disanthoideae (1/1)
                Subfamily: Hamamelidoideae (23/77)
                Subfamily: Rhodoleioideae (1/1)
                Subfamily: Symingtonioideae (1/2)
                Subfamily: Liquidambaroideae (2/10)
            Family: Platanaceae (C; 1/10)
    Order: Casuarinales (C; 2/70)
            Family: Casuarinaceae (C; 2/70)
    Order: Fagales (C; 14/1,240)
            Family: Fagaceae (C; 8/1,079)
                Subfamily: Fagoideae (2/45)
                Subfamily: Castaneoideae (4/434)
                Subfamily: Quercoideae (2/600)
            Family: Betulaceae (C; 6/157)
                Subfamily: Betuloideae (2/95)
                Subfamily: Coryloideae (4/62)

Superorder: Rosiflorae (C; 369/6,310)

    Order: Rosales (C; 331/5,790)
        Suborder: Rosineae (C; 120/2,430)
            Family: Rosaceae (C; 100/2,000)
                Subfamily: Spiraeoideae (incl. *Lyonothamnus*)
                Subfamily: Quillajeoideae (2/8)
                Subfamily: Rosoideae
                Subfamily: Pyroideae (incl. *Lindleya, Vauquelinia*)
                Subfamily: Prunoideae (incl. *Exochorda*)
                Subfamily: Neuradoideae (3/10)
            Family: Chrysobalanaceae (C; 17/420)
            Family: Crossosomataceae (incl. *Forsellesia*) (C; 3/11)
        Suborder: Saxifragineae (C; 182/2,800)
            Family: Crassulaceae (C; 35/1,500)
                Subfamily: Sedoideae (incl. Sempervivieae, Echeverieae)
                Subfamily: Cotyledonoideae (incl. Kalanchoeae)
                Subfamily: Crassuloideae
            Family: Cephalotaceae (C; 1/1)
            Family: Saxifragaceae (C; 78/810)
                Subfamily: Tetracarpoideae (1/1)
                Subfamily: Saxifragoideae (incl. *Francoa, Penthorum*) (34/180)
                Subfamily: Iteoideae (2/17)
                Subfamily: Ribesioideae (1/150)
                Subfamily: Brexioideae (excl. *Ixerba*)
                Subfamily: Pterostemonoideae (1/2)
                Subfamily: Vahlioideae (1/5)
                Subfamily: Eremosynoideae (1/1)
                Subfamily: Hydrangeoideae (incl. *Kirengeshoma, Philadelphus*)
                                                        (17/250)
                Subfamily: Escallonioideae (incl. *Corokia, Griselinia, Ixerba,
                                  Phyllonoma*) (10/165)
                Subfamily: Alseuosmioideae (2/11)

   Subfamily: Montinioideae (incl. *Grevea, Kaliphora, Melanophylla*)
                               (4/13)
    Subfamily: Columellioideae (1/4)
  Family: Parnassiaceae (B; 1/50)
  Family: Stylidiaceae (B; 6/170)
    Subfamily: Donatioideae (1/2)
    Subfamily: Stylidioideae (5/166)
  Family: Droseraceae (C; 4/110)
  Family: Greyiaceae (C; 1/3)
  Family: Podostemaceae (B; 50/140)
    Subfamily: Tristichoideae (5/10)
    Subfamily: Podostemoideae (45/130)
  Family: Diapensiaceae (B; 6/20)
 Suborder: Cunoniineae (C; 29/480)
  Family: Cunoniaceae (C; 20/350)
  Family: Baueraceae (C; 1/3)
  Family: Davidsoniaceae (D; 1/1)
  Family: Brunelliaceae (C; 1/52)
  Family: Eucryphiaceae (C; 1/6)
  Family: Staphyleaceae (C; 5/60)
    Subfamily: Staphyleoideae (3/55)
    Subfamily: Tapiscioideae (2/5)
Order: Pittosporales (C; 38/520)
 Suborder: Buxineae (incl. Daphniphyllineae) (B; 7/120)
  Family: Buxaceae (excl. *Simmondsia*) (C; 5/100)
    Subfamily: Buxoideae (4/100)
    Subfamily: Styloceratoideae (1/3)
  Family: Daphniphyllaceae (C; 1/9)
  Family: Balanopaceae (C; 1/0)
 Suborder: Pittosporineae (C; 14/285)
  Family: Pittosporaceae (C; 10/240)
  Family: Byblidaceae (C; 1/2)
  Family: Tremandraceae (C; 3/43)
 Suborder: Brunineae (B; 17/110)
  Family: Roridulaceae (C; 1/2)
  Family: Bruniaceae (C; 12/75)
  Family: Geissolomataceae (C; 1/1)
  Family: Grubbiaceae (C; 1/3)
  Family: Myrothamnaceae (C; 1/2)
  Family: Hydrostachyaceae (C; 1/30)

Superorder: Myrtiflorae (C; 463/8,115)

Order: Myrtales (C; 463/8,115)
  Family: Lythraceae (C; 27/460)
    Subfamily: Lythroideae (24/50)
    Subfamily: Sonneratioideae (2/7)
    Subfamily: Punicoideae (1/2)
  Family: Oliniaceae (C; 1/10)
  Family: Penaeaceae (C; 5/25)
  Family: Trapaceae (C; 1/prob. 1 or 3)
  Family: Melastomataceae (C; 247/3,370)
    Subfamily: Melastomatoideae (237/2,950)
    Subfamily: Astronioideae (3/50)
    Subfamily: Memecyloideae (4/360)
    Subfamily: Crypteronioideae (incl. *Axindandra, Dactylocladus*)
                         (3/10)

```
       Family:    Combretaceae (C; 20/600)
          Subfamily:  Strephonematoideae
          Subfamily:  Combretoideae
       Family:    Onagraceae (C; 18/650)
       Family:    Myrtaceae (C; 144/3,000)
          Subfamily:  Psiloxyloideae (1/1)
          Subfamily:  Leptospermoideae (incl. Heteropyxis, Kania)
          Subfamily:  Myrtoideae (incl. Osbornia)


Superorder:  Gentianiflorae (C; 1,765/21,230)

   Order:  Oleales (C; 32/610)
       Family:    Salvadoraceae (B/ 3/12)
       Family:    Oleaceae (C; 29/600)
          Subfamily:  Jasminoideae
          Subfamily:  Oleoideae
          Subfamily:  Nyctanthoideae (1/2)
   Order:  Gentianales (C; 975/11,340)
       Family:    Loganiaceae (C; 24/550)
          Subfamily:  Loganioideae (18/480)
          Subfamily:  Retzioideae (1/1)
          Subfamily:  Desfontainioideae (4/70)
          Subfamily:  Plocospermatoideae (1/1)
       Family:    Buddlejaceae (C; 10/150)
       Family:    Rubiaceae (C; 500/6,000)
          Subfamily:  Cinchonoideae
          Subfamily:  Guettardoideae
          Subfamily:  Ixoroideae
          Subfamily:  Rubioideae (incl. Theligonum)
          Subfamily:  Hillioideae
          Subfamily:  Henriquezioideae (2/13)
       Family:    Apocynaceae (C; 355/3,700)
          Subfamily:  Plumerioideae
          Subfamily:  Cerberoideae
          Subfamily:  Apocynoideae
          Subfamily:  Periplocoideae (45/200)
          Subfamily:  Secamonoideae
          Subfamily:  Asclepiadoideae
       Family:    Gentianaceae (C; 81/900)
          Subfamily:  Gentianoideae (80/900)
          Subfamily:  Saccifolioideae (1/1)
       Family:    Menyanthaceae (C; 5/40)
   Order:  Bignoniales (C; 762/9,280)
       Family:    Bignoniaceae (incl. Paulownieae) (C; 120/650)
       Family:    Pedaliaceae (C; 12/75)
       Family:    Martyniaceae (C; 4/16)
       Family:    Myoporaceae (incl. Leucophylleae) (C; 6/150)
       Family:    Scrophulariaceae (incl. Jerdonia, excl. Capraria, Leucophyll
                                                           (C; 216/3,030)
          Subfamily:  Scrophularioideae (incl. Selagineae)
          Subfamily:  Rhinanthoideae (incl. Globulareae)
          Subfamily:  Orobanchoideae (13/180)
       Family:    Plantaginaceae (C; 3/270)
       Family:    Lentibulariaceae (C; 4/170)
       Family:    Acanthaceae (C; 256/2,770)
          Subfamily:  Nelsonioideae (5/15)
```

            Subfamily: Mendoncioideae (2/60)
            Subfamily: Acanthoideae
            Subfamily: Ruellioideae
        Family: Gesneriaceae (C; 120/2,000)
            Subfamily: Cyrtandroideae
            Subfamily: Gesnerioideae

Superorder Lamiiflorae (C; 420/9,380)

    Order: Lamiales (C; 285/6,720)
        Family: Verbenaceae (C; 104/3,200)
            Subfamily: Viticoideae
            Subfamily: Verbenoideae
            Subfamily: Chloanthoideae (14/90)
            Subfamily: Lithophytoideae (1/1)
            Subfamily: Phrymatoideae (1/2)
            Subfamily: Caryopteridoideae (6/40)
            Subfamily: Stilboideae (5/12)
            Subfamily: Symphoematoideae (3/34)
            Subfamily: Avicennioideae (1/15)
        Family: Callitrichaceae (C; 1/25)
        Family: Lamiaceae (C; 180/3,500)
            Subfamily: Prostantheroideae (6/?)
            Subfamily: Ajugoideae (9/?)
            Subfamily: Prasiodeae (6/?)
            Subfamily: Ocimoideae
            Subfamily: Catopherioideae (1/3)
            Subfamily: Lavanduloideae (1/26)
            Subfamily: Rosmarinoideae (1/1)
            Subfamily: Scutellarioideae (2/201)
            Subfamily: Lamioideae (Stachyoideae)
            Subfamily: Tetrachondroideae (1/2)
    Order: Boraginales (B; 135/2,660)
        Family: Hydrophyllaceae (C; 18/250)
        Family: Boraginaceae (C; 114/2,400)
            Subfamily: Cordioideae
            Subfamily: Ehretioideae
            Subfamily: Heliotropoideae (incl. Ixorhea) (5/430)
            Subfamily: Boraginioideae (95/1,570)
            Subfamily: Wellstedioideae (1/2)
        Family: Lennoaceae (C; 3/5)

Superorder: Solaniflorae (B; 1,249/7,490)

    Order: Solanales (C; 166/5,160)
      Suborder: Solanineae (C; 149/4,830)
        Family: Solanaceae (C; 90/3,000)
            Subfamily: Solanoideae (55/2,400)
            Subfamily: Destroideae (26/485)
            Subfamily: Nolanoideae (2/85)
            Subfamily: Sclerophylacoideae (1/12)
            Subfamily: Duckeodendroideae (1/1)
            Subfamily: Goetzeoideae (5/7)
        Family: Convolvulaceae (C; 59/1,830)
            Subfamily: Humbertioideae (1/1)

                Subfamily: Dichondroideae (2/9)
                Subfamily: Convolvuloideae (55/1,650)
                Subfamily: Cuscutoideae (1/170)
        Suborder: Polemoniineae (C; 16/320)
            Family: Polemoniaceae (incl. *Cobaea*) (16/320)
        Suborder: Fouquieriineae (B; 1/11)
            Family: Fouquieriaceae (1/11)
    Order: Campanulales (B; 83/2,330)
            Family: Pentaphragmataceae (C; 1/30)
            Family: Campanulaceae (C; 66/2,000)
                Subfamily: Campanuloideae (34/820)
                Subfamily: Cyphioideae (incl. only *Cyphia*) (1/50)
                Subfamily: Lobelioideae (incl. *Cyphocarpus*) (30/1,130)
                Subfamily: Sphenocleoideae (1/2)
            Family: Goodeniaceae (B; 16/300)
                Subfamily: Goodenioideae
                Subfamily: Dampieroideae
                Subfamily: Brunonioideae (1/1)

Superorder: Corniflorae (C; 434/5,960)

   Order: Cornales (C; 50/1,120)
        Suborder: Rhizophoineae (B; 16/120)
            Family: Rhizophoraceae (16/120)
                Subfamily: Anisophylleoideae (4/36)
                Subfamily: Rhizophoroideae (12/84)
        Suborder: Vitineae (B; 13/730)
            Family: Vitaceae (13/730)
                Subfamily: Vitoideae (12/700)
                Subfamily: Leeoideae (1/34)
        Suborder: Haloragineae (B; 11/153)
            Family: Haloragaceae (C; 9/100)
            Family: Gunneraceae (B; 1/50)
            Family: Hippuridaceae (B; 1/3)
        Suborder: Cornineae (C; 10/120)
            Family: Nyssaceae (C; 3/7)
                Subfamily: Davidioideae (1/1)
                Subfamily: Nyssoideae (2/6)
            Family: Cornaceae (C; 4/72)
                Subfamily: Mastixioideae (1/25)
                Subfamily: Curtisioideae (1/1)
                Subfamily: Cornoideae (2/46)
            Family: Alangiaceae (C; 1/19)
            Family: Garryaceae (C; 1/12)
            Family: Aucubaceae (C; 1/3)
   Order: Araliales (C; 243/3,610)
            Family: Helwingiaceae (C; 1/5)
            Family: Torricelliaceae (C; 1/3)
            Family: Araliaceae (C; 340/3,600)
                Subfamily: Aralioideae (incl. Apiaceae) (65/750)
                Subfamily: Hydrocotyloideae
                Subfamily: Saniculoideae
                Subfamily: Apioideae
   Order: Dipsacales (C; 42/1,230)
            Family: Caprifoliaceae (C; 14/490)
                Subfamily: Caprifolioideae (12/450)
                Subfamily: Sambucoideae (incl. *Viburnum*) (2/40)

Family: Adoxaceae (C; 1/1)
Family: Valerianaceae (C; 13/400)
Family: Dipsacaceae (incl. *Morina, Triplostegia*) (C; 10/300)
Family: Calyceraceae (C; 4/40)

Superorder: Asteriflorae (C; 1,150/21,300)

Order: Asterales (1,150/21,300)
    Family: Asteraceae (1,150/21,300)
        Subfamily: Cichorioideae (Lactucoideae   (324/7,760)
            Tribe: Mutisieae (89/975)
            Tribe: Vernonieae (incl. Liabeae, Trichopira) (70/1,640)
            Tribe: Cichorieae (Lactuceae) (70/2,300)
            Tribe: Cardueae (incl. Carlinae, Echinopeae, Echinopsideae,
                  Eremothamneae, Gunelieae) (79/2,660)
            Tribe: Arctotideae (excl. *Ursinia*) (16/200)
        Subfamily: Asteroideae (Helianthoideae ((825/13,510)
            Tribe: Heliantheae (incl. Arnica, Bahiinae, Gaillardiinae, Tagetinae)
                                    (226/2,400)
            Tribe: Eupatorieae (60/2,000)
            Tribe: Astereae (135/2,500)
            Tribe: Inuleae (180/2,100)
            Tribe: Anthemideae (incl. Ursiniinae) (102/1,400)
            Tribe: Senecioneae (incl. Blennospermatinae) (114/3,000)
            Tribe: Calenduleae (8/110)

        Subclass: MONOCOTYLEDONEAE (LILIIDAE)

Superorder: Liliiflorae (C; 1,185/25,450)

Order: Liliales (C; 1,185/25,450)

    Suborder: Liliineae (C; 367/7,520)
        Family: Liliaceae (C; 345/6,460)
            Subfamily: Melanthioideae (24/140)
            Subfamily: Herrerioideae (3/9)
            Subfamily: Asphodeloideae (58/1,000)
            Subfamily: Dracaenoideae (7/820)
            Subfamily: Xanthorrhoeoideae (incl. *Hanguana*) (9/68)
            Subfamily: Wurmbaeoideae (13/165)
            Subfamily: Lilioideae (14/425)
            Subfamily: Scilloideae (33/600)
            Subfamily: Allioideae (30/600)
            Subfamily: Alstreomerioideae (4/200)
            Subfamily: Ixiolirioideae (2/3)
            Subfamily: Amaryllidoideae (63/860)
            Subfamily: Agavoideae (10/390)
            Subfamily: Hypoxidoideae (5/140)
            Subfamily: Haemodoroideae (22/120)
            Subfamily: Cyanastroideae (1/7)
            Subfamily: Asparagoideae (30/500)
            Subfamily: Ophiopogonoideae (3/18)
            Subfamily: Aletroideae (2/13)
            Subfamily: Luzuriagoideae (8/10)
            Subfamily: Smilacoideae (4/375)
        Family: Velloziaceae (C; 6/269)

            Subfamily: Barbacenioideae (2/39)
            Subfamily: Vellozioideae (4/230)
         Family: Stemonaceae (C; 3/30)
         Family: Dioscoreaceae (C; 10/750)
         Family: Trichopodaceae (C; 1/1)
         Family: Taccaceae (C; 2/10)
      Suborder: Iridineae (C; 83/930)
         Family: Iridaceae (C; 61/800)
            Subfamily: Iridoideae (60/800)
            Subfamily: Geosiridoideae (1/1)
         Family: Burmanniaceae (C; 22/130)
            Subfamily: Burmannioideae
            Subfamily: Thismioideae
            Subfamily: Corsioideae (2/9)
      Suborder: Orchidineae (C; 735/17,000)
         Family: Orchidaceae (C; 735/17,000)
            Subfamily: Apostasioideae (3/20)
            Subfamily: Cypripedioideae
            Subfamily: Spiranthoideae (incl. Neottieae)
            Subfamily: Orchidoideae
            Subfamily: Epidendroideae
            Subfamily: Vandoideae

Superorder: Triuridiflorae (B; 7/80)

   Order: Triuridales (7/80)
         Family: Triuridaceae (incl. *Peltophyllum*) (7/80)

Superorder: Alismatiflorae (C; 56/430)

   Order: Alismatales (C; 34/180)
         Family: Butomaceae (C; 1/1)
         Family: Alismataceae (C; 17/100)
            Subfamily: Limnocharitoideae (4/7)
            Subfamily: Alismatoideae (13/90)
         Family: Hydrocharitaceae (C; 16/80)
            Subfamily: Hydrocharitoideae (4/11)
            Subfamily: Vallisnerioideae (10/64)
            Subfamily: Thalassioideae (1/2)
            Subfamily: Halophiloideae (1/4)
   Order: Zosterales (C; 21/200)
      Suborder: Aponogetonineae (C; 1/30)
         Family: Aponogetonaceae (1/30)
      Suborder: Potamogetonineae (C; 18/150)
         Family: Juncaginaceae (C; 6/26)
            Subfamily: Scheuchzerioideae (1/1)
            Subfamily: Juncaginoideae (incl. *Lilaea*) (5/25)
         Family: Potamogetonaceae (incl. *Ruppia*) (C; 2/100)
         Family: Posidoniaceae (C; 1/3)
         Family: Zannichelliaceae (C; 4/7)
         Family: Cymodoceaceae (C; 5/16)
      Suborder: Zosterineae (C; 2/18)
         Family: Zosteraceae (C; 2/18)
   Order: Najadales (C; 1/50)
         Family: Najadaceae (C; 1/50)

Superorder: Areciflorae (C; 225/3,660)

   Order: Arecales (C; 211/2,780)
      Family: Arecaceae (C; 212/2,780)
         Subfamily: Coryphoideae (32/322)
         Subfamily: Phoenicoideae (1/17)
         Subfamily: Borassoideae (6/56)
         Subfamily: Lepidocaryoideae (22/664)
         Subfamily: Nypoideae (1/1)
         Subfamily: Carytoideae (3/35)
         Subfamily: Chaemaedoroideae (11/180)
         Subfamily: Arecoideae (incl. Cocosoideae) (131/1,490)
         Subfamily: Phytelephantoideae (4/15)
   Order: Cyclanthales (C; 11/180)
      Family: Cyclanthaceae (11/180)
         Subfamily: Carludovicoideae (10/179)
         Subfamily: Cyclanthoideae (1/1)
   Order: Pandanales (C; 3/700)
      Family: Pandanaceae (3/700)
         Subfamily: Pandanoideae (2/600)
         Subfamily: Freycinetioideae (1/100)

Superorder: Ariflorae (C; 104/1,530)

   Order: Arales (C; 104/1,530)
      Family: Araceae (C; 100/1,500)
         Subfamily: Acorodeae (2/4)
         Subfamily: Pothoideae (9/450)
         Subfamily: Monsteroideae (11/175)
         Subfamily: Calloideae (4/5)
         Subfamily: Lasioideae (16/150)
         Subfamily: Philodendroideae (16/350)
         Subfamily: Calocasioideae (14/160)
         Subfamily: Aroideae (27/200)
         Subfamily: Pistioideae (1/1)
      Family: Lemnaceae (C; 4/28)
         Subfamily: Lemnoideae (2/14)
         Subfamily: Wolffioideae (2/14)

Superorder: Typhiflorae (B; 2/30)

   Order: Typhales (2/30)
      Family: Typhaceae (C; 2/30)
         Subfamily: Sparganioideae (1/20)
         Subfamily: Typhoideae (1/10)

Superorder: Commeliniflorae (C; 1,027/20,360)

   Order: Commelinales (C; 940/18,870)
     Suborder: Bromeliineae (C; 64/2,350)
       Family: Bromeliaceae (C; 44,2000)
          Subfamily: Navioideae
          Subfamily: Pitcairnioideae
          Subfamily: Tillandsioideae
          Subfamily: Bromelioideae

   Family: Rapateaceae (C; 16/80)
    Subfamily: Saxofridericioideae
    Subfamily: Rapateoideae
   Family: Xyridaceae (incl. Abolbodaceae) (C; 4/270)
  Suborder: Pontederiineae (C; 11/35)
   Family: Pontederiaceae (C; 7/30)
   Family: Philydraceae (C; 4/5)
  Suborder: Juncineae (C; 100/4,400)
   Family: Juncaceae (C; 10/400)
    Subfamily: Thurnioideae (1/3)
    Subfamily: Juncoideae (9/400)
   Family: Cyperaceae (C; 90/4,000)
    Subfamily: Cyperoideae
    Subfamily: Caricoideae
    Subfamily: Mapanioideae
  Suborder: Commelinineae (C; 39/510)
   Family: Commelinaceae (incl. *Cartonema*) (C; 38/500)
   Family: Mayacaceae (C; 1/10)
  Suborder: Eriocaulineae (C; 12/1,200)
   Family: Eriocaulaceae (C; 12/1,200)
    Subfamily: Eriocauloideae
    Subfamily: Paepalanthoideae
  Suborder: Flagellariineae (C; 42/365)
   Family: Flagellariaceae (excl. *Hanguana*) (C; 2/5)
    Subfamily: Flagellarioideae (1/3)
    Subfamily: Joinvilleoideae (1/2)
   Family: Restionaceae (C; 36/330)
    Subfamily: Restionoideae (36/330)
    Subfamily: Anarthrioideae (1/5)
    Subfamily: Ecdeiocoleaceae (C; 2/3)
   Family: Centrolepidaceae (excl. Hydatellaceae) (C; 4/30)
  Suborder: Poineae (C; 670/10,000)
   Family: Poaceae (C; 670/10,000)
    Subfamily: Bambusoideae (incl. *Anomochloa, Pariana, Streptochaete*)
    Subfamily: Oryzoideae
    Subfamily: Arundinoideae (incl. Stipeae)
    Subfamily: Centothecoideae (incl. *Calderonella*)
    Subfamily: Micrairoideae (1/2)
    Subfamily: Pooideae (Festucoideae)
    Subfamily: Eragrostoideae
    Subfamily: Panicoideae
Order: Zingiberales (C; 87/1,490)
   Family: Musaceae (C; 2/42)
   Family: Strelitziaceae (C; 3/7)
   Family: Heliconiaceae (C; 1/80)
   Family: Lowiaceae (C; 1/2)
   Family: Zingiberaceae (C; 45/700)
   Family: Costaceae (C; 4/200)
   Family: Cannaceae (C; 1/55)
   Family: Marantaceae (C; 30/400)

TAXA INCERTAE SEDIS

Barbeya Schwinf. (1/1) Removed from Urticales.
Capraria L. (1/4) Usually placed in Scrophulariaceae.
Corynocarpus J.R. & G. Forst (1/5) Possibly in Rosales.

Dialypetalanthus Kuhlm. (1/1) Probably in or near the Rubiaceae.
Heteranthia Nees et Mart. (1/1) Probably in or near Solanaceae.
Hoplestigma Pierre (1/2) Sometimes placed in Boraginales.
Hydatellaceae Hamann (2/7) Probably either Liliiflorae or Commeliniflorae.
Physena Nor. ex Thou. (1/2) Sometimes treated in Capparaceae.
Setchellanthus T.S. Brandegee (1/1) Sometimes treated in Capparaceae.